CHAPMAN & HALL/CRC APPLIED MATHEMATICS
AND NONLINEAR SCIENCE SERIES

INTRODUCTION TO FUZZY SYSTEMS

Published Titles

Geometric Sturmian Theory of Nonlinear Parabolic Equations and Applications,
   Victor A. Galaktionov
Mathematical Methods in Physics and Engineering with Mathematica,
   Ferdinand F. Cap
Introduction to Fuzzy Systems, Guanrong Chen and Trung Tat Pham
Introduction to Partial Differential Equations with MATLAB®, Matthew P. Coleman
Optimal Estimation of Dynamic Systems, John L. Crassidis and John L. Junkins

Forthcoming Titles

Mathematical Theory of Quantum Computation, Goong Chen and Zijian Diao
Stochastic Partial Differential Equations, Pao-Liu Chow

CHAPMAN & HALL/CRC APPLIED MATHEMATICS
AND NONLINEAR SCIENCE SERIES

INTRODUCTION TO FUZZY SYSTEMS

Guanrong Chen
City University of Hong Kong
Kowloon, Hong Kong, P R China

Trung Tat Pham
DIcentral Corporation
Houston, Texas, USA

Chapman & Hall/CRC
Taylor & Francis Group

Boca Raton   London   New York

Published in 2006 by
Chapman & Hall/CRC
Taylor & Francis Group
6000 Broken Sound Parkway NW, Suite 300
Boca Raton, FL 33487-2742

International Standard Book Number-10: 1-58488-531-9 (Hardcover)
International Standard Book Number-13: 978-1-58488-531-3 (Hardcover)

Library of Congress Cataloging-in-Publication Data

Catalog record is available from the Library of Congress

Taylor & Francis Group
is the Academic Division of Informa plc.

Visit the Taylor & Francis Web site at
http://www.taylorandfrancis.com

and the CRC Press Web site at
http://www.crcpress.com

Preface

The word "fuzzy" is no longer fuzzy to many engineers today. Introduced in the earlier 1970s, fuzzy systems and fuzzy control methodologies as an emerging technology targeting industrial applications has added a promising new dimension to the existing domain of conventional control systems engineering. It is now a common belief that when a complex physical system does not provide a set of differential or difference equations as a precise or reasonably accurate mathematical model, particularly when the system description requires certain human experience in linguistic terms, fuzzy systems and fuzzy control theories have some salient features and distinguishing merits over many other approaches.

Fuzzy control methods and algorithms, including many specialized software and hardware available on the markets, may be classified as one type of intelligent control. This is because fuzzy systems modeling, analysis, and control incorporate a certain amount of human knowledge into its components such as fuzzy sets, fuzzy logic, and fuzzy rule bases. Using human expertise in system modeling and controller design is not only advantageous but often necessary. Classical controllers design has already incorporated human knowledge and skills; for instance, what type of controller to use and how to determine the controller structure and parameters largely depend on the decision and preference of the designer, especially when multiple choices are possible. The fuzzy control technology provides just one more choice for this consideration; it has the intention to be an alternative, by no means a simple replacement, of the existing control techniques such as classical control and other intelligent control methods (e.g., neural networks, expert systems, etc.). Together, they supply systems and control engineers with a more complete toolbox to deal with the complex, dynamic, and uncertain real world. Fuzzy control technology is one of the many tools in this toolbox that is developed not only for elegant mathematical theories, but more importantly, for many practical problems with various technical challenges.

Compared with various conventional approaches, fuzzy control utilizes more information from domain experts and yet relies less on mathematical modeling about a physical system.

On the one hand, fuzzy control theory can be quite heuristic and somewhat ad hoc. This sometimes is preferable or even desirable, particularly when low-cost and easy operations are required where mathematical rigor is not the main concern. There are many examples of this kind in industrial applications, for which fuzzy sets and fuzzy logic are easy to use. Within this context, determining a fuzzy set or a fuzzy rule base seems to be somewhat subjective, where human knowledge about the underlying physical system comes into play. However, this may not require more human knowledge than selecting a suitable mathematical model in a deterministic control approach, where the following questions are often being asked beforehand: "Should one use a linear or a nonlinear model?" "If linear, what's the order or dimension; if nonlinear, what type of nonlinearity?" "What kind of optimality criterion should be used to measure the performance?" "What kind of norm should be used to measure the robustness?" It is also not much more subjective than choosing a suitable distribution function in the stochastic control approach, where the following questions are frequently being asked: "Gaussian or non-Gaussian noise?" "White noise or just unknown but bounded uncertainty?" Although some of these questions can be answered on the basis of statistical analysis of available empirical data in classical control systems, the same is true for establishing an initial fuzzy rule base in fuzzy control systems.

On the other hand, fuzzy control theory can be rigorous and fuzzy controllers can have precise formulations with analytic structures and guaranteed closed-loop system stability and performance specifications, if such characteristics are intended. In this direction, the ultimate objective of the current fuzzy systems and fuzzy control research is appealing: the fuzzy control system technology is moving toward a solid foundation as part of the modern control theory. The trend of a rigorous approach to fuzzy control, starting from the mid-1980s, has produced many exciting and promising results. For instance, some analytic structures of fuzzy controllers, particularly fuzzy PID controllers, and their relationship with corresponding conventional controllers are much better understood today. Numerous analysis and design methods have been well developed. As a consequence, the existing analytical control theory has made the fuzzy control systems practice safer, more efficient, and more cost-effective.

This textbook represents a continuing effort in the pursuit of analytic theory and rigorous design for fuzzy control systems. More specifically, the basic notion of fuzzy mathematics (Zadeh fuzzy set theory, fuzzy membership functions, interval and fuzzy number arithmetic operations) is first studied. Consequently, in a comparison with the classical two-valued logic, the fundamental concept of fuzzy logic is introduced. Some real-world applications of fuzzy logic will then be discussed, just after two chapters of studies, revealing some practical flavors of fuzzy logic. This is then followed by the basic fuzzy systems theory (Mamdani and Takagi-Sugeno modeling, along with parameter estimation and system identification) and fuzzy control theory. Here, fuzzy control theory is introduced, mainly based on the well-known classical Proportional-Integral-Derivative (PID) controllers theory and design methods. In particular, fuzzy PID controllers are studied in greater detail. These controllers have precise analytic structures, with rigorous analysis and guaranteed closed-loop system stability; they are comparable and also compatible with the classical PID controllers. To that end, a new notion of verb-based fuzzy-logic control theory is briefly described.

The primary purpose of this course is to provide some rather systematic training for systems and control majors, both senior undergraduate and first-year graduate students, and to familiarize them with some fundamental mathematical theory and design methodology in fuzzy control systems. The authors have tried to make this book self-contained, so that no preliminary knowledge of fuzzy mathematics and fuzzy control systems theory is needed to understand the materials presented in this textbook. Although it is assumed that the students are aware of some very basic classical set theory and classical control systems theory, the fundamentals of these subjects are briefly reviewed throughout the text for their convenience.

Some common terminology in the field of fuzzy control systems has become quite standard today. Therefore, as a textbook written in a classical style, the authors have taken the liberty to omit some personal and specialized names such as "TS fuzzy model" and "t-norm," to name just a couple. One reason is that too many names have to be given to too many items in doing so, which will distract the readers' attention in their reading. Nevertheless, closely related references are given at the end of the book for crediting and for one's further searching. Also, an * in the Table of Contents indicates those relatively advanced materials that are beyond the basic scope of the present text; they are used only for the reader's further studies of the subject and can be omitted in regular teaching.

This textbook is a significantly modified version of the authors' book "Introduction to Fuzzy Sets, Fuzzy Logic, and Fuzzy Control Systems" (CRC Press, 2001), which has been used by the authors for a graduate course in the City University of Hong Kong since its publication, prior to which the authors' Lecture Notes had been taught for several years in the University of Houston at Texas of USA. This new book differs from the original one in many aspects. First of all, two new chapters, Chapters 3 and 7, have been added, with emphasis on some real applications of fuzzy logic and some new development of the verb-based fuzzy control theory and methodology. To keep the book within a modest size and make it more readable to new comers, some advanced topics on adaptive fuzzy control and high-level applications of fuzzy logic, presented in Chapters 6 and 7 of the original book, have been removed. Secondly, almost all chapters in the original book have been simplified, keeping the most fundamental contents and aiming at a more elementary textbook that can be easily used for teaching of the subject. Last but not least, more practical examples and review problems have been added or revised, with problem solutions provided at the end of the book, which should benefit both class-room teaching and self-studying.

In the preparation of this new textbook, the authors received some suggestions and typo-corrections on the previous book from their students each year. In particular, Dr. Tao Yang provided the materials of Chapter 7 on verb-based fuzzy control theory, which has quite significantly enhanced the contents of the book.

It is the authors' hope that students will benefit from this textbook in obtaining some relatively comprehensive knowledge about fuzzy control systems theory which, together with their mathematical foundations, can in a way better prepare them for the rapidly developing applied control technologies emerging from the modern industries.

The Authors

Table of Contents

CHAPTER 1

Fuzzy Set Theory

The classical set theory is built on the fundamental concept of "set." An individual is either a member or not a member of a specified set in question. A sharp, crisp, and unambiguous distinction exists between a member and a nonmember for any well-defined "set" in this classical theory of mathematics, and there is a very precise and clear boundary indicating whether or not an individual belongs to the set. When one is asked the question "Is this individual a member of that set?" The answer is either "yes" or "no."

The classical concept of "set" holds for both the deterministic and the stochastic cases. In probability and statistics, one may ask, "What is the probability of this individual being a member of that set?" Although an answer to this question could be like "The probability for this individual to be a member of that set is 90%," the final outcome (i.e., conclusion) is still either "it is" or "it is not" a member of that set. Here, the chance for one to make a correct prediction as "it is a member of that set" is 90%, which does not mean that it has 90% membership in the set and, meanwhile, it possesses 10% non-membership of the same set. In other words, in the classical set theory, it is not permissible for an individual to be partially in a set and also partially not in the same set at the same time. Thus, many real-world application problems cannot be described and, as a result, cannot be solved by the classical set theory, including all those involving elements with only partial membership of a set. On the contrary, fuzzy set theory accepts partial memberships; therefore, in a sense it generalizes the classical set theory to some extent.

In order to introduce the concept of fuzzy sets, the elementary set theory in classical mathematics is first reviewed. It will be seen that the fuzzy set theory is a very natural extension of the classical set theory, and is also a rigorous mathematical notion.

I. CLASSICAL SET THEORY

Let S be a nonempty set, called the *universe set* below, consisting of all possible elements of interest within a particular context.

Each of these elements is called a *member*, or an *element*, of S. A union of some (finite or infinitely many) members of S is called a *subset* of S.

To indicate that a member s of S belongs to a subset S of $\mathbf{S}$, one usually writes

$s \in S$,

but if s is not a member of S,

$s \notin S$.

To indicate that S is a subset of $\mathbf{S}$, one writes

$S \subset \mathbf{S}$.

Usually, this notation implies that S is a strictly proper subset of $\mathbf{S}$ in the sense that there is at least one member $x \in \mathbf{S}$ but $x \notin S$. If it can be either $S \subset \mathbf{S}$ or $S = \mathbf{S}$, then one writes

$S \subseteq \mathbf{S}$.

An empty subset is denoted by $\varnothing$. A subset of certain members that all have properties $P_1, \ldots, P_n$ will be denoted by a capital letter, say A, as

$A = \{\, a \mid a \text{ has properties } P_1, \ldots, P_n \,\}$.

A subset $A \subseteq \mathbf{R}^n$ is said to be *convex* if

$$x = \begin{bmatrix} x_1 \\ \vdots \\ x_n \end{bmatrix} \in A \qquad \text{and} \qquad y = \begin{bmatrix} y_1 \\ \vdots \\ y_n \end{bmatrix} \in A$$

implies

$$\lambda x + (1 - \lambda)y \in A \qquad \text{for any } \lambda \in [0,1].$$

Let A and B be two subsets. If every member of A is also a member of B, i.e., if $a \in A$ implies $a \in B$, then A is said to be a *subset* of B, and is denoted $A \subset B$.

If both $A \subset B$ and $B \subset A$ are true, then they are *equal*, for which it is indicated by $A = B$. If it can be either $A \subset B$ or $A = B$, then it is denoted $A \subseteq B$. Therefore, $A \subset B$ is equivalent to both $A \subseteq B$ and $A \neq B$.

The *difference* of two subsets A and B is defined by

$$A - B = \{ \, c \mid c \in A \text{ and } c \notin B \, \}.$$

In particular, if $A = S$ is the universe set, then $S - B$ is called the *complement* of B, and is denoted by $\overline{B}$, namely,

$$\overline{B} = S - B.$$

Clearly,

$$\overline{\overline{B}} = B, \quad \overline{S} = \varnothing, \text{ and } \overline{\varnothing} = S.$$

Let $r \in R$ be a real number and A be a subset of R. Then, the *multiplication* of r and A is defined to be the set

$$rA = \{ \, ra \mid a \in A \, \}.$$

The *union* of two subsets A and B is defined as

$$A \cup B = B \cup A = \{ \, c \mid c \in A \text{ or } c \in B \, \}.$$

Thus, one always has

$$A \cup S = S, \quad A \cup \varnothing = A, \text{ and } \quad A \cup \overline{A} = S.$$

The *intersection* of two subsets A and B is defined by

$$A \cap B = B \cap A = \{ \, c \mid c \in A \text{ and } c \in B \, \}.$$

Clearly,

$$A \cap S = A, \quad A \cap \varnothing = \varnothing, \text{ and } \quad A \cap \overline{A} = \varnothing.$$

Moreover, two subsets A and B are said to be *disjoint* if

$$A \cap B = \varnothing.$$

Basic properties of the classical set theory are summarized in Table 1.1, where $A \subseteq S$ and $B \subseteq S$.

For any set A, the *characteristic function* of A is defined by

$$X_A(x) = \begin{cases} 1 & \text{if } x \in A, \\ 0 & \text{if } x \in A. \end{cases} \qquad (1.1)$$

For any two sets A and B and for any element $x \in S$, one has

$$X_{A \cup B}(x) = \max\{ X_A(x), X_B(x) \},$$

$$X_{A \cap B}(x) = \min\{ X_A(x), X_B(x) \},$$

$$X_{\bar{A}}(x) = 1 - X_A(x).$$

Table 1.1 Properties of Classical Set Operations

Involutive law	$\bar{\bar{A}} = A$
Commutative law	$A \cup B = B \cup A$ $A \cap B = B \cap A$
Associative law	$(A \cup B) \cup C = A \cup (B \cup C)$ $(A \cap B) \cap C = A \cap (B \cap C)$
Distributive law	$A \cap (B \cup C) = (A \cap B) \cup (A \cap C)$ $A \cup (B \cap C) = (A \cup B) \cap (A \cup C)$ $A \cup A = A$ $A \cap A = A$ $A \cup (A \cap B) = A$ $A \cap (A \cup B) = A$ $A \cup (\bar{A} \cap B) = A \cup B$ $A \cap (\bar{A} \cup B) = A \cap B$ $A \cup S = S$ $A \cap \emptyset = \emptyset$ $A \cup \emptyset = A$ $A \cap S = A$ $A \cap \bar{A} = \emptyset$ $A \cup \bar{A} = S$
DeMorgan's law	$\overline{A \cap B} = \bar{A} \cup \bar{B}$ $\overline{A \cup B} = \bar{A} \cap \bar{B}$

II. FUZZY SET THEORY

The following simple example serves as an introduction to the concept of fuzzy sets.

Example 1.1. Let S be the set of all human beings and consider its subset

$$S_f = \{\, s \in S \mid s \text{ is old}\, \}.$$

Then, S_f is a "fuzzy set" because the property "old" is not well defined in the sense of classical mathematics and cannot be precisely measured: given a person who is 40-year old, it is not clear if this person belongs to the set S_f, and even if so, it is still unclear whether or not a person of age 39 also belongs to the same set.

In classical set theory, one may draw a line at the exact age of 40. As a result, a person who is exactly 40 years old belongs to the set and is considered to be "old," but another person of one-day-less than 40 years old will not be considered "old." This distinction is mathematically correct, but practically unreasonable.

To do mathematics, one nevertheless has to precisely define the subset S_f. For this purpose, one needs to quantify the concept "old," so as to characterize the subset S_f in a precise and rigorous way.

Instead of a sharp cut at the exact age of 40, let's say, one would like to describe the concept "old" by the curve shown in Figure 1.1 (a), using common sense, where the only ones who are considered to be "absolutely old" are those 120-year old or older, and the only people who are considered to be "absolutely young" are those newborns. Meanwhile, all the other people are old as well as young at the same time, with different degrees of oldness and youngness depending on their actual ages.

For example, a person 40 years old is considered to be "old" with "degree 0.5" and at the same time also "young" with "degree 0.5" according to the measuring curve that has been chosen for use. One cannot exclude this person from the subset S_f, nor include him completely.

Thus, the curve in Figure 1.1 (a) establishes a mathematical measure for the oldness of a human being. The curve shown in Figure 1.1 (a) is called a *membership function* associated with the subset S_f. It is a generalization of the classical characteristic function X_{S_f} defined by (1.1), which can only be used to conclude that a person either "is" or "is not" a member of the subset S_f.

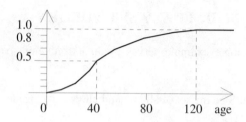

Figure 1.1 (a) A fuzzy membership function for "oldness"

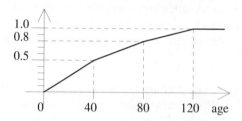

Figure 1.1 (b) Another fuzzy membership function for "oldness"

One may also use the piecewise linear membership function shown in Figure 1.1 (b) to describe the same concept of oldness for the same subset S_f, depending on whichever is more meaningful and more convenient for the application under consideration. Clearly, both membership functions are reasonable and acceptable in common sense.

In general, a fuzzy membership function can have various shapes, as shown in Figure 1.2, chosen by the user based on the nature of the application in consideration.

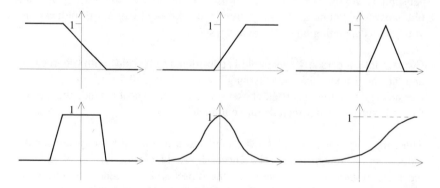

Figure 1.2 Various shapes of commonly used membership functions

Next, consider again the subset

$$S_f = \{ s \in S \mid s \text{ is old} \}.$$

Suppose that a membership function associated with it, say the one shown in Figure 1.1 (a), has been chosen for use. Then, this subset S_f, along with the chosen membership function, denoted by $\mu_{S_f}(s)$ with $s \in S_f$, is called a *fuzzy set*.

It is now clear that a fuzzy set consists of two components: a regular set and a membership function associated with it. This is different from the classical set theory, where all sets (and subsets) actually share the same (and the only) simple membership function: the two-valued characteristic function X_{S_f} defined by (1.1), which is not even being mentioned because in such simple cases it is not necessary to do so.

To familiarize this new concept, consider one more example in the following.

Example 1.2. Let S be the set of real numbers and let

$$S_f = \{ s \in S \mid s \text{ is positive and large} \}.$$

This set, S_f, is not well-defined in the sense of classical set theory because, although the statement "s is positive" is precise, the statement "s is large" is vague ("fuzzy").

However, if one introduces a membership function, reasonable and meaningful in the present discussion, to characterize or measure the property "large," then the fuzzy set S_f, associated with this membership function $\mu_{S_f}(s)$, is well defined.

Here, the membership function shown in Figure 1.3 may be chosen to use.

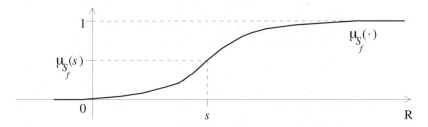

Figure 1.3 A membership function for a positive and large real number

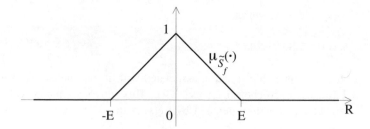

Figure 1.4 A membership function for a real number of small value

This membership function is quantified by

$$\mu_{S_f}(s) = \begin{cases} 0 & \text{if } s \le 0, \\ 1 - e^{-s} & \text{if } s > 0, \end{cases}$$

which is reasonable for describing a positive and "large" real number in common sense. Depending on how large is considered to be large in the application at hand, some other functions may be chosen instead. This is the user's choice. It is similar to the case when one is doing least-squares data fitting; if he believes a straight-line is simple and reasonable to use for a particular data set, then he selects it to use.

Similarly, a membership function for the set

$$\tilde{S}_f = \{ s \in S \mid |s| \text{ is small} \}$$

may be chosen to be the one shown in Figure 1.4, where the cutting edge E is determined by the user according to his knowledge and preference about the concerned application.

Summary of general features of fuzzy membership functions:

All membership functions discussed above have been normalized to have maximum value 1, as usual, since 1 = 100% describes a full membership and is convenient to use.

Although a membership function is a nonnegative-valued function, it differs from the probability density functions in that the area under the curve of a membership function does not have to be equal to unity (in fact, it can be any value between 0 and ∞, including 0 and ∞). Moreover, a membership function does not have to be continuous, or integrable.

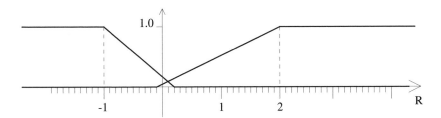

Figure 1.5 $s = 0.1$ is both "positive large" and "negative small"

Another distinction between the fuzzy set theory and the classical one is that a member of a fuzzy set may assume two or more membership values, and these membership values can even be conflicting. For example, if the two membership functions shown in Figure 1.5 are used to measure "positive and large" and "negative and small," respectively, then a member $s = 0.1$ has the first membership value 0.095 and the second 0.080: they do not sum up to 0.175, nor cancel out to be 0.015. Moreover, these two concepts are conflicting: s is positive and in the meantime also negative, with different degrees of correctness to be so. This situation is just like someone who is old and simultaneously young, which classical mathematics cannot accept. Such a vague and conflicting description of a fuzzy set is acceptable by fuzzy mathematics: in fact, it turns out to be very useful in describing and solving many real-world application problems where conflicting conditions are not uncommon. More importantly, the use of conflicting membership functions will not cause any logical or mathematical problems in the consequence, provided that a correct approach is taken carefully. Such a correct approach does exist; that is the *fuzzy set theory* to be further studied in the following sections of this chapter.

III. INTERVAL ARITHMETIC

Recall that a *fuzzy set* consists of two parts: a *set* defined in the ordinary sense and a *membership* function defined on the set, and this membership function is also defined in the ordinary sense.

Now, some fundamental properties and operation rules pertaining to a special yet important kind of sets – *intervals* – are first studied, which will be needed in the sequel.

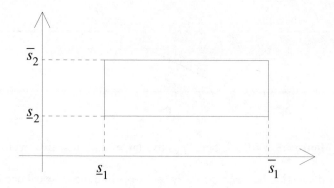

Figure 1.6 An interval of confidence in the two-dimensional case

A. Some Fundamental Concepts

The concern here is the situation where the value of a member s of a set is *uncertain* but bounded:

$$\underline{s} \leq s \leq \bar{s} \, ,$$

where $[\, \underline{s} \, , \, \bar{s} \,] \subset R$ is called the *interval of confidence* about the values of s. Only closed intervals in the form of $[\, \underline{s} \, , \, \bar{s} \,]$ are considered in this text, not those like $(\, \underline{s} \, , \, \bar{s} \,]$, $[\, \underline{s} \, , \, \bar{s} \,)$, $(\, \underline{s} \, , \, \bar{s} \,)$, except perhaps $[\, \underline{s} \, , \, \infty)$ and $(-\infty, \bar{s} \,]$ in some special cases.

A special case is, when $\underline{s} = \bar{s}$, it becomes $[\, \underline{s} \, , \, \bar{s} \,] = [s \, , s \,] = s$.

In the two-dimensional case, an interval of confidence is sometimes called a *region of confidence*, as shown in Figure 1.6.

Definition 1.1

(a) ***Equality***: Two intervals $[\, \underline{s}_1, \, \bar{s}_1 \,]$ and $[\, \underline{s}_2, \, \bar{s}_2 \,]$ are said to be equal if and only if $\underline{s}_1 = \underline{s}_2$ and $\bar{s}_1 = \bar{s}_2$:

$$[\underline{s}_1, \bar{s}_1 \,] \; = \; [\, \underline{s}_2, \bar{s}_2 \,]$$

(b) ***Intersection***: The *intersection* of two intervals $[\, \underline{s}_1, \, \bar{s}_1 \,]$ and $[\, \underline{s}_2, \, \bar{s}_2 \,]$ is

$$[\underline{s}_1, \bar{s}_1 \,] \cap [\, \underline{s}_2, \bar{s}_2 \,] = [\, \max\{\underline{s}_1, \underline{s}_2\} \, , \, \min\{ \bar{s}_1, \bar{s}_2 \} \,]$$

Note: $[\underline{s}_1, \bar{s}_1 \,] \cap [\, \underline{s}_2, \bar{s}_2 \,] = \varnothing$ if and only if $\underline{s}_1 > \bar{s}_2$ or $\underline{s}_2 > \bar{s}_1$.

(c) **Union**: The *union* of two intervals $[\underline{s}_1, \overline{s}_1]$ and $[\underline{s}_2, \overline{s}_2]$ is

$$[\underline{s}_1, \overline{s}_1] \cup [\underline{s}_2, \overline{s}_2] = [\min\{\underline{s}_1, \underline{s}_2\}, \max\{\overline{s}_1, \overline{s}_2\}],$$

provided that $[\underline{s}_1, \overline{s}_1] \cap [\underline{s}_2, \overline{s}_2] \neq \varnothing$. Otherwise, it is undefined (since the result is not an interval).

(d) **Inequality**: Interval $[\underline{s}_1, \overline{s}_1]$ is said to be *less than* (resp., *greater than*) interval $[\underline{s}_2, \overline{s}_2]$, denoted by

$$[\underline{s}_1, \overline{s}_1] < [\underline{s}_2, \overline{s}_2] \qquad (\text{resp.}, [\underline{s}_1, \overline{s}_1] > [\underline{s}_2, \overline{s}_2])$$

if and only if $\overline{s}_1 < \underline{s}_2$ (resp., $\underline{s}_1 > \overline{s}_2$). Otherwise, they cannot be compared. Note that the relations $\leq$ and $\geq$ are not defined for intervals.

(e) **Inclusion**: The interval $[\underline{s}_1, \overline{s}_1]$ is *being included* in the interval $[\underline{s}_2, \overline{s}_2]$ if and only if both $\underline{s}_2 \leq \underline{s}_1$ and $\overline{s}_1 \leq \overline{s}_2$:

$$[\underline{s}_1, \overline{s}_1] \subseteq [\underline{s}_2, \overline{s}_2]$$

Example 1.3. For three intervals, $S_1 = [-1,0]$, $S_2 = [-1,2]$, and $S_3 = [2,10]$:

$$S_1 \cap S_2 = [-1,0] \cap [-1,2] = [-1,0],$$

$$S_1 \cap S_3 = [-1,0] \cap [2,10] = \varnothing,$$

$$S_2 \cap S_3 = [-1,2] \cap [2,10] = [2,2] = 2,$$

$$S_1 \cup S_2 = [-1,0] \cup [-1,2] = [-1,2],$$

$$S_1 \cup S_3 = [-1,0] \cup [2,10] = \text{undefined},$$

$$S_2 \cup S_3 = [-1,2] \cup [2,10] = [-1,10],$$

$$S_1 = [-1,0] < [2,10] = S_3,$$

$$S_1 = [-1,0] \subset [-1,2] = S_2.$$

B. Interval Arithmetic

Let $[\underline{s}, \overline{s}]$, $[\underline{s}_1, \overline{s}_1]$, and $[\underline{s}_2, \overline{s}_2]$ be intervals.

Definition 1.2 Interval Arithmetic

(1) *Addition*:

$$[\,\underline{s}_1,\,\bar{s}_1\,] + [\,\underline{s}_2,\,\bar{s}_2\,] = [\,\underline{s}_1 + \underline{s}_2,\,\bar{s}_1 + \bar{s}_2\,].$$

(2) *Subtraction*:

$$[\,\underline{s}_1,\,\bar{s}_1\,] - [\,\underline{s}_2,\,\bar{s}_2\,] = [\,\underline{s}_1 - \bar{s}_2,\,\bar{s}_1 - \underline{s}_2\,].$$

(3) *Reciprocal:*

If $0 \notin [\,\underline{s},\,\bar{s}\,]$ then $[\,\underline{s},\,\bar{s}\,]^{-1} = [\,1/\bar{s},\,1/\underline{s}\,]$;

if $0 \in [\,\underline{s},\,\bar{s}\,]$ then $[\,\underline{s},\,\bar{s}\,]^{-1}$ is undefined.

(4) *Multiplication*:

$$[\,\underline{s}_1,\,\bar{s}_1\,] \cdot [\,\underline{s}_2,\,\bar{s}_2\,] = [\,\underline{p},\,\bar{p}\,],$$

where

$$\underline{p} = \min\{\,\underline{s}_1\,\underline{s}_2,\,\underline{s}_1\,\bar{s}_2,\,\bar{s}_1\,\underline{s}_2,\,\bar{s}_1\,\bar{s}_2\,\},$$

$$\bar{p} = \max\{\,\underline{s}_1\,\underline{s}_2,\,\underline{s}_1\,\bar{s}_2,\,\bar{s}_1\,\underline{s}_2,\,\bar{s}_1\,\bar{s}_2\,\}.$$

(5) *Division*:

$$[\,\underline{s}_1,\,\bar{s}_1\,] / [\,\underline{s}_2,\,\bar{s}_2\,] = [\,\underline{s}_1,\,\bar{s}_1\,] \cdot [\,\underline{s}_2,\,\bar{s}_2\,]^{-1},$$

provided that $0 \notin [\,\underline{s}_2,\,\bar{s}_2\,]$.

(6) *Maximum*:

$$\max\{\,[\,\underline{s}_1,\,\bar{s}_1\,],\,[\,\underline{s}_2,\,\bar{s}_2\,]\} = [\,\underline{p},\,\bar{p}\,],$$

where

$$\underline{p} = \max\{\,\underline{s}_1,\,\underline{s}_2\,\},$$

$$\bar{p} = \max\{\,\bar{s}_1,\,\bar{s}_2\,\}.$$

(7) *Minimum*:

$$\min\{\,[\,\underline{s}_1,\,\bar{s}_1\,]\,,[\,\underline{s}_2,\,\bar{s}_2\,]\} = [\,\underline{p}\,,\,\bar{p}\,]\,,$$

where

$$\underline{p} \;=\; \min\{\,\underline{s}_1\,,\,\underline{s}_2\,\},$$

$$\bar{p} \;=\; \min\{\,\bar{s}_1,\,\bar{s}_2\,\}.$$

Remarks:

(a) Interval arithmetic intends to obtain an interval as the result of an operation such that the resulting interval contains all possible solutions. Therefore, these kinds of operational rules are defined in a conservative way, in the sense that they intend to make the resulting interval as large as necessary so as to avoid losing any true solution. For example, $[1,2]-[0,1]=[0,2]$ means that for any $a\in[1,2]$ and any $b\in[0,1]$, it is guaranteed that $a-b\in[0,2]$.

(b) This conservatism may produce some unusual results that could seem to be inconsistent with the ordinary numerical solutions. For instance, according to the subtraction rule (2), one has $[1,2]-[1,2]=[-1,1]\neq[0,0]$ $=0$. The result $[-1,1]$ here contains 0, but not only 0. The reason is that there can be other possible solutions: if one takes 1.5 from the first interval and 1.0 from the second, then the result is 0.5 rather than 0; and 0.5 is indeed in $[-1,1]$. Thus, an interval subtract itself is equal to zero (a point) only if this interval is itself a point (a trivial interval).

(c) Some general rules for the interval arithmetic:

For any interval Z,

$$Z-Z=0 \qquad \text{and} \qquad Z/Z=I=[1,1] \qquad (0\notin Z)$$

only if $Z=[z,z]$ is a point.

For any intervals X, Y, and Z,

$$X+Z=Y+Z\Rightarrow X=Y.$$

For any interval Z, with $0\in Z$,

$$Z^2 = Z \cdot Z = [\ \underline{z},\ \overline{z}\ ] \cdot [\ \underline{z},\ \overline{z}\ ] = [\ \underline{p},\ \overline{p}\ ],$$

where

$$\underline{p} \quad = \min\{\ \underline{z}^2,\ \underline{z}\,\overline{z},\ \overline{z}^{\,2}\ \} = \underline{z}\,\overline{z},$$

$$\overline{p} \quad = \max\{\ \underline{z}^2,\ \underline{z}\,\overline{z},\ \overline{z}^{\,2}\ \} = \max\{\ \underline{z}^2,\ \overline{z}^{\,2}\}.$$

(d) Every computational system has restrictions (e.g., the ordinary arithmetic does not allow dividing by zero). Interval arithmetic is no exception. Fortunately, not much interval arithmetic will be involved in fuzzy systems, fuzzy-logic-based decision making, and fuzzy control applications, at least not much are needed within the context of this text, therefore generally no confusion would arise about such "unusual" phenomena and rules. Although many more details exist in the mathematical literature about special interval arithmetic rules, so that conflicts can be avoided in a way similar to the problems caused by "dividing by zero" in classical mathematics, this issue will not be further discussed here.

Note that the interval operations of addition (+), subtraction (−), multiplication (·), and division (/) are all (set-variable and set-valued) *functions*: for three intervals, $X = [\ \underline{x},\ \overline{x}\ ]$, $Y = [\ \underline{y},\ \overline{y}\ ]$, and $Z = [\ \underline{z},\ \overline{z}\ ]$,

$$Z = f(X,Y) = X * Y, \qquad\qquad\qquad * \in \{+, -, \cdot, / \}$$

are continuous functions defined on intervals.

C. Algebraic Properties of Interval Arithmetic

Theorem 1.1 The addition and multiplication operations of intervals are *commutatitve* and *associative* but not *distributive*:

(1) $X + Y = Y + X$;

(2) $Z + (X + Y) = (Z + X) + Y$;

(3) $Z (X Y) = (Z X) Y$;

(4) $X Y = Y X$;

(5) $Z + 0 = 0 + Z = Z$ and $Z0 = 0Z = 0$, where $0 = [0,0]$;

(6) $ZI = IZ = Z$, where $I = [1,1]$;

(7) $Z(X + Y) \neq ZX + ZY$, except when:

    (a) $Z = [z,z]$ is a point; or

    (b) $X = Y = \mathbf{0}$; or

    (c) $xy \geq 0$ for all $x \in X$ and $y \in Y$.

In general, only the *subdistributive* law holds:

$$Z(X + Y) \subseteq ZX + ZY.$$

Example 1.4. Let

$$Z = [1,2], X = I = [1,1], Y = -I = [-1,-1].$$

Then

$$Z(X + Y) = [1,2](I - I) = [1,2]\cdot\mathbf{0} = \mathbf{0};$$

$$ZX + ZY = [1,2] \cdot [1,1] + [1,2] \cdot [-1,-1] = [-1,1] \supset \mathbf{0}.$$

A more general rule for interval arithmetic operations is the following fundamental law of *monotonic inclusion*, established in the mathematical literature.

Theorem 1.2 Let X_1, X_2, Y_1, and Y_2 be intervals such that

$$X_1 \subseteq Y_1 \quad \text{and} \quad X_2 \subseteq Y_2.$$

Then, for all operations $* \in \{ +, -, \cdot, / \}$,

$$X_1 * X_2 \subseteq Y_1 * Y_2.$$

Corollary 1.1 Let X and Y be intervals with $x \in X$ and $y \in Y$. Then, for any operation $* \in \{ +, -, \cdot, / \}$,

$$x * y \in X * Y$$

Finally, consider the problem of solving the *interval equation*

$$AX = B,$$

where A and B are both given intervals with $0 \notin A$, and X is to be determined.

Theorem 1.3 Let X be a solution of the interval equation

$$AX = B, \ 0 \notin A \ .$$

Then, $X \subseteq B / A$.

Example 1.5

Consider the interval equation $AX=B$, with

$$A = \begin{bmatrix} \dfrac{1}{2} & -\dfrac{1}{2} \\ \dfrac{1}{2} & \dfrac{1}{2} \end{bmatrix} \quad \text{and} \quad B = \begin{bmatrix} [0,1] \\ [0,1] \end{bmatrix} .$$

Then

$$B/A = \begin{bmatrix} \dfrac{1}{2} & -\dfrac{1}{2} \\ \dfrac{1}{2} & \dfrac{1}{2} \end{bmatrix}^{-1} \begin{bmatrix} [0,1] \\ [0,1] \end{bmatrix} = \begin{bmatrix} 1 & 1 \\ -1 & 1 \end{bmatrix} \begin{bmatrix} [0,1] \\ [0,1] \end{bmatrix} = \begin{bmatrix} [0,2] \\ [-1,1] \end{bmatrix} .$$

It is clear that $\begin{bmatrix} 0 \\ 1 \end{bmatrix} \in B / A$ but it is not a solution of the interval

equation $AX=B$, namely,

$$\begin{bmatrix} \dfrac{1}{2} & -\dfrac{1}{2} \\ \dfrac{1}{2} & \dfrac{1}{2} \end{bmatrix} \begin{bmatrix} 0 \\ 1 \end{bmatrix} \neq \begin{bmatrix} [0,1] \\ [0,1] \end{bmatrix} .$$

Therefore, $X \subset B / A$ but $X \neq B / A$.

All the above basic theoretical results are standard in interval mathematics. Hence, only the conclusions, but not their proofs, are provided here, as preliminaries.

D. Interval Evaluation

An ordinary real-variable and real-valued functions f: R→R can easily be extended to an interval-variable and interval-valued function f: **I**→**I**, where **I** is the family of intervals defined on R. Such extended functions include the following arithmetic functions:

$$Z = f(X,Y) = X * Y, \quad * \in \{ +, -, \cdot, / \},$$

where $X, Y, Z \in$ **I**.

Note that for any ordinary continuous function f: R→R and any interval $X \in$ **I**, the interval-variable and interval-valued function

$$f_{\mathbf{I}}(X) = \left[\min_{x \in X} f(x), \max_{x \in X} f(x) \right]$$

is also a continuous function.

Now, let $A_1, ..., A_m$ be intervals in **I**. For any interval $X \in$ **I**, one can further define a function, $f(x;a_1,...,a_m)$, $x \in X$, which depends on m parameters $a_k \in A_k$, $k = 1, 2,...,m$, by

$$f_{\mathbf{I}}(X;A_1,...,A_m) = \{ f(x;a_1,...,a_m) \mid x \in X, a_k \in A_k, 1 \le k \le m \}$$

$$= \left[\min_{\substack{x \in X \\ a_k \in A, 1 \le k \le m}} f(x,a_1,...,a_m), \max_{\substack{x \in X \\ a_k \in A, 1 \le k \le m}} f(x,a_1,...,a_m) \right]$$

Example 1.6

Consider the real-variable and real-valued function

$$f(x;a) = \frac{ax}{1-x}, \quad x \ne 1, \quad x \ne 0.$$

If $X = [2, 3]$ and $A = [0, 2]$ are intervals, with $x \in X$ and $a \in A$, then the interval expression of f is given by

$$f_{\mathbf{I}}(X;A) = \left\{ \frac{ax}{1-x} \;\middle|\; 2 \le x \le 3, \; 0 \le a \le 2 \right\}$$

$$= \left[\min_{\substack{2 \le x \le 3 \\ 0 \le a \le 2}} \frac{ax}{1-x}, \ \max_{\substack{2 \le x \le 3 \\ 0 \le a \le 2}} \frac{ax}{1-x} \right]$$

$$= [-4, 0].$$

The following result is important and useful. It states that all common interval arithmetic expressions have the *inclusion monotonic* property.

Theorem 1.4 Let $f : R^{n+m} \rightarrow R$ be a real-variable and real-valued continuous function with an arithmetic interval expression $f_I(X_1,...,X_n,A_1,...,A_m)$. Then, for all

$$X_k \subseteq Y_k, \quad k = 1,2,...,n \quad \text{and} \quad A_l \subseteq B_l, \quad l = 1,2, \ ... \ , m,$$

one has

$$f_I(X_1,...,X_n,A_1,...,A_m) \subseteq f_I(Y_1,...,Y_n,B_1,...,B_m) \ .$$

Example 1.7

Let $X = [0.2, 0.4]$ and $Y = [0.1, 0.5]$. Then $X \subset Y$.

(a) $X^{-1} = \dfrac{1}{[0.2, 0.4]} = [2.5, 5.0]$,

$Y^{-1} = \dfrac{1}{[0.1, 0.5]} = [2.0, 10.0]$,

$X^{-1} \subset Y^{-1}$.

(b) $1 - X = [1.0, 1.0] - [0.2, 0.4] = [0.6, 0.8]$,

$1 - Y = [1.0, 1.0] - [0.1, 0.5] = [0.5, 0.9]$,

$1 - X \subset 1 - Y$.

(c) $\dfrac{1}{1-X} = \dfrac{1}{[0.6, 0.8]} = [5/4, 5/3]$,

$\dfrac{1}{1-Y} = \dfrac{1}{[0.5, 0.9]} = [10/9, 2.0]$,

$$\frac{1}{1-X} \subset \frac{1}{1-Y}.$$

A very important issue in the evaluation of an interval expression is that the number of intervals involved in an interval expression should be reduced (whenever possible) before evaluating (i.e., computing) the expression, in order to obtain less-conservative lower and upper bounds of the resulting interval expression.

Example 1.8

Consider the function discussed in Example 1.6:

$$f(x;a) = \frac{ax}{1-x}, \quad x \neq 1, \ x \neq 0 \ .$$

If one rewrites it as

$$\tilde{f}(x;a) = \frac{a}{\dfrac{1}{x} - 1}, \quad x \neq 1, \ x \neq 0,$$

then, as a real-variable and real-valued function, $\tilde{f} \equiv f$.

In the case of intervals, for $x \in X = [2, 3]$ and $a \in A = [0, 2]$, according to Example 1.6, one has

$$f_I(X;A) = [-4, 0] \ .$$

On the other hand, one also has

$$\tilde{f}_I(X;A) = \left\{ \frac{a}{1/x - 1} \,\middle|\, 2 \leq x \leq 3, 0 \leq a \leq 2 \right\}$$

$$= \left[\min_{\substack{2 \leq x \leq 3 \\ 0 \leq a \leq 2}} \frac{a}{\dfrac{1}{x} - 1}, \ \max_{\substack{2 \leq x \leq 3 \\ 0 \leq a \leq 2}} \frac{a}{\dfrac{1}{x} - 1} \right] = [-4, 0] = f_I(X;A) \ .$$

However, if one formally performs interval arithmetic (rather than numerical minimization and maximization, as just did), then one obtains

$$f_{\mathbf{I}}(X;A) = \frac{[0,2] \cdot [2,3]}{1-[2,3]} = \frac{[0,6]}{[-2,-1]} = [0,6].[-1, -1/2] = [-6, 0]$$

and

$$\tilde{f}_{\mathbf{I}}(X;A) = \frac{[0,2]}{\dfrac{1}{[2,3]} -1} = \frac{[0,2]}{\left[\dfrac{1}{3},\dfrac{1}{2}\right] -1} = \frac{[0,2]}{\left[-\dfrac{2}{3},-\dfrac{1}{2}\right]}$$

$$= [0, 2] \cdot [-2, -3/2]$$

$$= [-4, 0].$$

Thus, $\tilde{f}_{\mathbf{I}}(X;A) \neq f_{\mathbf{I}}(X;A)$ but

$$[-4, 0] = \tilde{f}_{I}(X;A) \subseteq f_{\mathbf{I}}(X;A) = [-6, 0].$$

The reason is that formula $f_{\mathbf{I}}(X;A)$ has three intervals, but $\tilde{f}_{I}(X;A)$ has only two, so that the computational "errors" are accumulated more in the former; therefore, the result is more conservative.

Note that, this example has shown a fundamental principle of interval arithmetic: one should always try to reduce the number of intervals involved in evaluating an interval expression.

Example 1.9. Calculating the power series of an interval X:

$$f_{\mathbf{I}}(X) = I + X + X^2 + X^3 + \ldots + X^n .$$

The best way is to first reformulate it as

$$\tilde{f}_{I}(X) = I + X (I + X (I + \ldots + X (I + X) \ldots))$$

and then carry out the calculations. The reason is that one needs to apply $1 + 2 + 3 + \ldots + n = (\frac{1}{2})(n)(n+1)$ times of the interval X in computing $f_{\mathbf{I}}$ but only use n times of X in computing $\tilde{f}_{I}$, where $X(I+X) \subseteq X + X^2$ by Theorem 1.1 (7).

Example 1.10. Find the *best* (i.e., the smallest possible) interval solution of

$$\frac{X^2 -1}{X^2} , \quad \text{where} \quad X = [a,3], \ 0 < a < 3.$$

Solution:

$$\frac{X^2-1}{X^2} = 1-\frac{1}{X^2} = 1-\frac{1}{[a,3]^2} = 1-\frac{1}{[a^2,9]} = 1-\left[\frac{1}{9},\frac{1}{a^2}\right]$$

$$= \left[\frac{a^2-1}{a^2},\frac{8}{9}\right].$$

For comparison, a correct but not the best solution is:

$$\frac{X^2-1}{X^2} = \frac{[a,3]^2-1}{[a,3]^2} = [[a^2,9]-1]\cdot[a^2,9]^{-1}$$

$$= [a^2-1,8]\cdot\left[\frac{1}{9},\frac{1}{a^2}\right] = \left[\frac{a^2-1}{9},\frac{8}{a^2}\right],$$

which is bigger than the above (best) solution because $0 < a < 3$ therefore $a^2 < 9$.

IV. OPERATIONS ON FUZZY SETS

The interval arithmetic studied in the last section is now extended to fuzzy sets.

A. Fuzzy Sets and α Cuts

A fuzzy set, denoted S_f, is defined to be a set S together with an associate membership function $\mu_{S_f}(s)$, as shown in Figure 1.7.

Definition 1.3 The subsets S_α of S_f defined by

$$S_\alpha = \{ s \in S_f \mid \mu_{S_f}(s) \geq \alpha \}, \quad \text{for an} \quad \alpha \in [0,1]$$

is called an α-*cut* of the fuzzy set S_α.

The concept of α-cuts is visualized via Figure 1.8, where $S_{0.8}$ and $S_{0.6}$ are both α-cuts, but the latter is not a (single) interval, so the latter will not be used later.

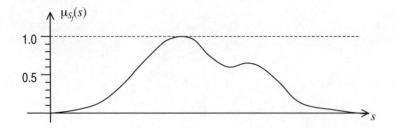

Figure 1.7 A fuzzy set is a set with an associate membership function

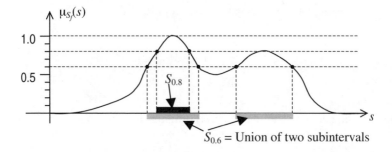

$S_{0.6}$ = Union of two subintervals

Figure 1.8 Two examples of α-cut

B. Arithmetic of Fuzzy Sets

A fuzzy set is called a *normal fuzzy set*, if its support is an interval, its membership function is convex, and it achieves the maximum number 1. Only normal fuzzy sets are considered in this text, so they are simply called *fuzzy sets*.

Recall that a function f defined on an interval A is *convex* if $x \in A$ and $y \in A$ implies

$$\lambda f(x) + (1 - \lambda) f(y) \leq f(\lambda x + (1 - \lambda) y) \quad \text{for any } \lambda \in [0,1].$$

For arithmetic of fuzzy sets, the following is a general rule, established in logic literature and adopted in this text.

General Rule Let X and Y be two fuzzy sets, with $Z \subseteq R$, and consider a two-variable function (e.g., $+$, $-$, $\cdot$,$/$, max, min):

$$F: X \times Y \to Z.$$

Let $\mu_X(x)$, $\mu_Y(y)$, and $\mu_Z(z)$ be their associate membership functions. Given $\mu_X(x)$ and $\mu_Y(y)$, define

$$\mu_Z(z) = \underset{z=F(x,y)}{\vee} \{ \mu_X(x) \wedge \mu_Y(y) \}.$$

where, for two real numbers s_1 and s_2,

$$s_1 \wedge s_2 = \min\{s_1, s_2\} \quad \text{and} \quad s_1 \vee s_2 = \max\{s_1, s_2\}.$$

Using the α-cut notation, this is equivalent to the following:

$$(Z)_\alpha = F((X)_\alpha, (Y)_\alpha)$$

$$= \{ z \in Z \mid z = F(x,y), x \in (X)_\alpha, y \in (Y)_\alpha \}.$$

The above computational rule and its α-cut notation are now illustrated by several examples.

Addition Let $z = F(x,y) = x + y$. Then

$$Z = \{ z \mid z = x + y, x \in X, y \in X \}$$

and

$$\mu_Z(z) = \underset{z=x+y}{\vee} \{ \mu_X(x) \wedge \mu_Y(y) \}.$$

In the α-cut notation:

$$(Z)_\alpha = F((X)_\alpha, (Y)_\alpha) = (X)_\alpha + (Y)_\alpha.$$

Example 1.11

Let x and y be such that

$$X = [-5,1], \qquad Y = [-5,12],$$

with associate membership functions shown in Figure 1.9:

$$\mu_X(x) = \begin{cases} \dfrac{x}{3} + \dfrac{5}{3}, & -5 \le x \le -2, \\[2mm] -\dfrac{x}{3} + \dfrac{1}{3}, & -2 \le x \le 1, \end{cases}$$

and

$$\mu_Y(y) \;=\; \begin{cases} 0 & -5 \leq y \leq -3, \\[4pt] \dfrac{y}{7} + \dfrac{3}{7} & -3 \leq y \leq 4, \\[4pt] -\dfrac{y}{8} + \dfrac{12}{8} & 4 \leq y \leq 12. \end{cases}$$

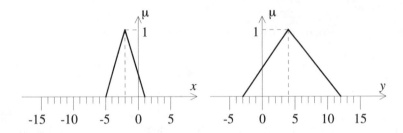

Figure 1.9 Two membership functions for addition

Then, by the general operation rule, one has

$$Z = X + Y = [-5,1] + [-5,12] = [-10,13]$$

and, by comparing $\mu_X(x)$ and $\mu_Y(y)$ point-wise, one obtains

$$\mu_Z(z) \;=\; \bigvee_{z=x+y} \{ \, \mu_X(x) \wedge \mu_Y(y) \, \}$$

$$= \begin{cases} 0, & -10 \leq z \leq -8, \\[4pt] \dfrac{z}{10} + \dfrac{8}{10}, & -8 \leq z \leq 2, \\[4pt] -\dfrac{z}{11} + \dfrac{13}{11}, & 2 \leq z \leq 13. \end{cases}$$

Here, it is clear that the general "sup" rule does not yield the explicit formulas easily. On the contrary, an explicit formula for $\mu_Z(z)$ can be easily obtained by using the equivalent α-cut operation as follows.

In the α-cut notation, for any α value, the α-cut of X is obtained by letting $\alpha = x/3 + 5/3$ and $\alpha = -x/3 + 1/3$, respectively, which give $x_1 = 3\alpha - 5$ and $x_2 = -3\alpha + 1$, as shown in (the enlarged) Figure 1.10.

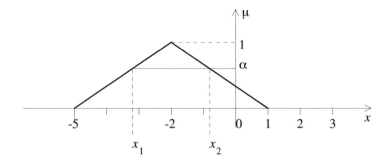

Figure 1.10 The α-cut of the membership function $\mu_Z(x)$

Hence, the projection interval is

$$(X)_\alpha = [\, x_1, x_2 \,] = [3\alpha - 5, -3\alpha + 1].$$

Similarly,

$$(Y)_\alpha = [7\alpha - 3, -8\alpha + 12],$$

so that

$$(Z)_\alpha = (X)_\alpha + (Y)_\alpha = [10\alpha - 8, -11\alpha + 13].$$

Taking $Z = [-10,13]$ into account, one finally arrives at

$$\mu_Z(z) = \begin{cases} 0, & -10 \le z \le -8, \\ \dfrac{z}{10} + \dfrac{8}{10}, & -8 \le z \le 2, \\ -\dfrac{z}{11} + \dfrac{13}{11}, & 2 \le z \le 13. \end{cases}$$

<u>Subtraction</u> Let $z = F(x,y) = x - y$. Then

$$Z = \{\, z \mid z = x - y,\ x \in X,\ y \in X \,\}$$

and

$$\mu_Z(z) = \underset{z = x - y}{\vee} \{\, \mu_X(x) \wedge \mu_Y(y) \,\}.$$

In the α-cut notation:

$$(Z)_\alpha = F((X)_\alpha, (Y)_\alpha) = (X)_\alpha - (Y)_\alpha.$$

Example 1.12

Let X and Y be such that

$$X = [0,20], \quad Y = [0,10],$$

with the membership functions

$$\mu_X(x) = \begin{cases} 0, & 0 \le x \le 7, \\ \dfrac{x}{7} - 1, & 7 \le x \le 14, \\ -\dfrac{x}{5} + \dfrac{19}{5}, & 14 \le x \le 19, \\ 0, & 19 \le x \le 20, \end{cases}$$

and

$$\mu_Y(y) = \begin{cases} 0, & 0 \le y \le 3, \\ \dfrac{y}{2} - \dfrac{3}{2}, & 3 \le y \le 5, \\ -\dfrac{y}{5} + 2, & 5 \le y \le 10. \end{cases}$$

Setting $z_1 = 10\alpha - 8$ and $z_2 = -11\alpha + 13$ gives $\alpha = z_1/10 + 8/10$ and $\alpha = -z_2/11 + 12/11$, which yield the membership function shown in Figure 1.11.

Then, one obtains, via the interval arithmetic, that

$$Z = X - Y = [-10,20],$$

with

$$\mu_Z(z) = \begin{cases} 0, & -10 \le z \le -3, \\ \dfrac{z}{12} + \dfrac{3}{12}, & -3 \le z \le 9, \\ -\dfrac{z}{7} + \dfrac{16}{7}, & 9 \le z \le 16, \\ 0, & 16 \le z \le 20. \end{cases}$$

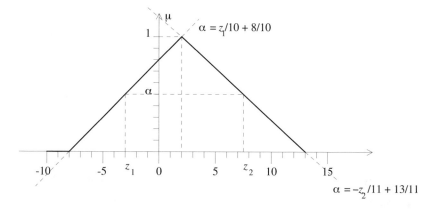

Figure 1.11 The resulting membership function of Example 1.11

In the α-cut notation, for any α value, the α-cut of X is obtained by letting $\alpha = x/7 - 1$ and $\alpha = -x/5 + 19/5$, respectively, which give $x_1 = 7\alpha + 7$ and $x_2 = -5\alpha + 19$. Hence, the projection interval is

$$(X)_\alpha = [\, x_1, x_2 \,] = [7\alpha + 7, \, -5\alpha + 19].$$

Similarly,

$$(Y)_\alpha = [2\alpha + 3, \, -5\alpha + 10],$$

so that

$$(Z)_\alpha = (X)_\alpha + (Y)_\alpha = [12\alpha - 3, \, -7\alpha + 16].$$

Setting $z_1 = 12\alpha - 3$ and $z_2 = -7\alpha + 16$ gives $\alpha = z_1/12 + 1/4$ and $\alpha = -z_2/7 + 16/7$, which yield the membership function shown in Figure 1.12.

<u>Multiplication</u> Let $z = F(x,y) = x \cdot y$. Then

$$Z = \{\, z \mid z = x\, y, x \in X, y \in Y \,\}$$

and

$$\mu_Z(z) = \bigvee_{z = x \cdot y} \{\, \mu_X(x) \wedge \mu_Y(y) \,\}.$$

In the α-cut notation:

$$(Z)_\alpha = F(\, (X)_\alpha, (Y)_\alpha \,) = (X)_\alpha \cdot (Y)_\alpha.$$

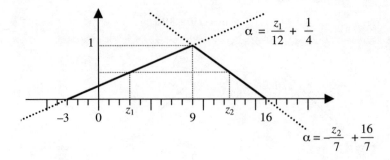

Figure 1.12 The resulting membership function of Example 1.12

Example 1.13

Let X and Y be such that

$$X = [2,5], \qquad Y = [3,6],$$

with the membership functions as shown in Figure 1.13:

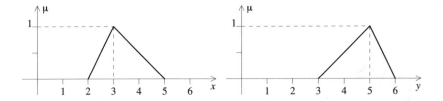

Figure 1.13 Two membership functions for Example 1.13

$$\mu_X(x) = \begin{cases} x - 2, & 2 \le x \le 3, \\ -\dfrac{x}{2} + \dfrac{5}{2}, & 3 \le x \le 5, \end{cases}$$

and

$$\mu_Y(y) = \begin{cases} \dfrac{y}{2} - \dfrac{3}{2}, & 3 \le y \le 5, \\ -y + 6, & 5 \le y \le 6. \end{cases}$$

In the α-cut notation, for any α value, letting

$$\alpha = x - 2 \qquad \text{and} \qquad \alpha = -\frac{x}{2} + \frac{5}{2}$$

gives

$$x_1 = \alpha + 2 \qquad \text{and} \qquad x_2 = -2\alpha + 5,$$

so that

$$(S_X)_\alpha = [\, \alpha + 2, \, -2\alpha + 5 \,].$$

Similarly,

$$(S_Y)_\alpha = [\, 2\alpha + 3, \, -\alpha + 6 \,].$$

It then follows that

$$S_Z = [6, 30]$$

and

$$\begin{aligned}
(S_Z)_\alpha &= (S_X)_\alpha \cdot (S_Y)_\alpha \\
&= [\alpha + 2, -2\alpha + 5] \cdot [\, 2\alpha + 3, -\alpha + 6 \,] \\
&= [\underline{p}(\alpha), \overline{p}(\alpha)\,],
\end{aligned}$$

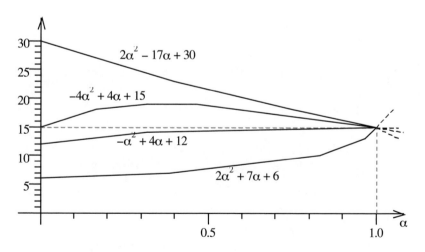

Figure 1.14 The intermediate membership functions of Example 1.13

where

$$\underline{p}(\alpha) = min\{2\alpha^2+7\alpha+6, -\alpha^2+4\alpha+12, -4\alpha^2+4\alpha+15, 2\alpha^2-17\alpha+30\},$$

$$\overline{p}(\alpha) = max\{2\alpha^2 + 7\alpha + 6, -\alpha^2 + 4\alpha + 12, -4\alpha^2 + 4\alpha + 15, 2\alpha^2 - 17\alpha + 30\},$$

with the curves shown in Figure 1.14.

Hence, one has

$$\underline{p}(\alpha) = 2\alpha^2 + 7\alpha + 6 \quad \text{and} \quad \overline{p}(\alpha) = 2\alpha^2 - 17\alpha + 30,$$

so that

$$(Z)_\alpha = [\underline{p}(\alpha), \overline{p}(\alpha)] = [2\alpha^2 + 7\alpha + 6, 2\alpha^2 - 17\alpha + 30].$$

Let, moreover,

$$z_1 = 2\alpha^2 + 7\alpha + 6 \quad \text{and} \quad z_2 = 2\alpha^2 - 17\alpha + 30.$$

Then, one can solve them for α, subject to $0 \le \alpha \le 1$, and obtain

$$\alpha = \frac{-7 + \sqrt{1 + 8z_1}}{4} \quad \text{or} \quad \alpha = \frac{17 - \sqrt{49 + 8z_2}}{4}.$$

Consequently,

$$\mu_Z(z) = \begin{cases} \dfrac{-7 + \sqrt{1 + 8z}}{4}, & 6 \le z \le 15, \\ \dfrac{17 - \sqrt{49 + 8z}}{4}, & 15 \le z \le 30, \end{cases}$$

as shown in Figure 1.15.

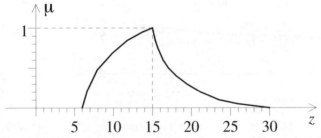

Figure 1.15 The resulting membership function of Example 1.13

<u>Division</u> Let $z = F(x,y) = \dfrac{x}{y}$. Then

$$Z = \{\, z \mid z = x/y, \; x \in X, \, y \in Y \,\}$$

and

$$\mu_Z(z) = \bigvee_{z = x/y} \{\, \mu_X(x) \wedge \mu_Y(y) \,\}.$$

In the α-cut notation:

$$(Z)_\alpha = F(\, (X)_\alpha, \, (Y)_\alpha \,) = \frac{(X)_\alpha}{(Y)_\alpha}.$$

Example 1.14

Let x and y be such that

$$X = [18,33], \qquad\qquad Y = [5,8],$$

with the membership functions

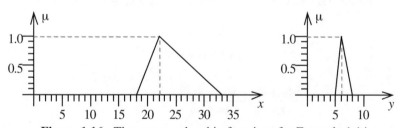

Figure 1.16 The two membership functions for Example 1.14

$$\mu_X(x) = \begin{cases} \dfrac{x}{4} - \dfrac{18}{4}, & 18 \le x \le 22, \\[2mm] -\dfrac{x}{11} + 3, & 22 \le x \le 33, \end{cases}$$

and

$$\mu_Y(y) = \begin{cases} y - 5, & 5 \le y \le 6, \\[2mm] -\dfrac{y}{2} + 4, & 6 \le y \le 8, \end{cases}$$

as shown in Figure 1.16.

In the α-cut notation, let

$$\alpha = \frac{x}{4} - \frac{18}{4} \qquad \text{and} \qquad \alpha = -\frac{x}{11} + 3.$$

One has

$$x_1 = 4\alpha + 18 \qquad \text{and} \qquad x_2 = -11\alpha + 33,$$

so that

$$(X)_\alpha = [\, 4\alpha + 18, \, -11\alpha + 33 \,].$$

Similarly,

$$(Y)_\alpha = [\, \alpha + 5, \, -2\alpha + 8 \,].$$

Hence,

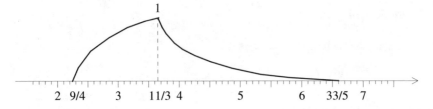

Figure 1.17 The resulting membership function of Example 1.14

$$(Z)_\alpha \quad = \frac{(X)_\alpha}{(Y)_\alpha} = \frac{[4\alpha + 18, -11\alpha + 33]}{[\alpha + 5, -2\alpha + 8]}$$

$$= \left[\, \frac{4\alpha + 18}{-2\alpha + 8}, \frac{-11\alpha + 33}{\alpha + 5} \,\right].$$

Next, letting

$$z_1 = \frac{4\alpha + 18}{-2\alpha + 8} \qquad \text{and} \qquad z_2 = \frac{-11\alpha + 33}{\alpha + 5},$$

gives

$$\alpha = \frac{8z_1 - 18}{2z_1 + 4} \qquad \text{and} \qquad \alpha = \frac{-5z_2 + 33}{z_2 + 11},$$

so that

$$\mu_Z(z) = \begin{cases} \dfrac{8z-18}{2z+4}, & \dfrac{9}{4} \le z \le \dfrac{11}{3}, \\[2ex] \dfrac{-5z+33}{z+11}, & \dfrac{11}{3} \le z \le \dfrac{33}{5}, \end{cases}$$

as shown in Figure 1.17.

Finally in this section, minimum and maximum operations on fuzzy sets are discussed.

Minimum Let X and Y be two fuzzy sets with

$$X = [\,a, b\,] \quad \text{and} \quad Y = [\,c, d\,],$$

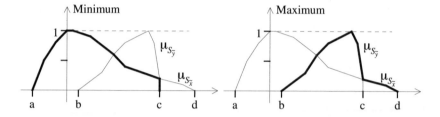

Figure 1.18 Fuzzy minimum and maximum membership functions

and with membership functions $\mu_X(x)$ and $\mu_Y(y)$, respectively. For any α value, denote

$$(X)_\alpha = [\,\underline{x}(\alpha), \bar{x}(\alpha)\,] \quad \text{and} \quad (Y)_\alpha = [\,\underline{y}(\alpha), \bar{y}(\alpha)\,].$$

The *fuzzy minimum* of x and y is defined to be $z = x \wedge y$, with

$$Z = [\,\min\{\,a, c\,\}, \min\{\,b, d\,\}\,]$$

and

$$\mu_Z(z) = \bigvee_{z=\min\{x,y\}} \{\,\mu_X(x) \wedge \mu_Y(y)\,\}$$

$$= \bigvee_{z=\min\{x,y\}} \{\,\mu_X(x) \wedge \mu_Y(y)\,\}$$

In the α-cut notation:

$$(Z)_\alpha = (X)_\alpha \wedge (Y)_\alpha = [\,\underline{x}(\alpha) \wedge \underline{y}(\alpha), \bar{x}(\alpha) \wedge \bar{y}(\alpha)\,].$$

This operation is visualized via Figure 1.18. An example will be given below, together with the maximum operation, to show how the result is obtained.

Maximum The *fuzzy maximum* of x and y is defined by $z = x \lor y$, with

$$Z = [\max\{a, c\}, \max\{b, d\}]$$

and

$$\mu_Z(z) = \bigvee_{z=\max\{x,y\}} \{\mu_X(x) \land \mu_Y(y)\}.$$

In the α-cut notation:

$$(Z)_\alpha = (X)_\alpha \lor (Y)_\alpha = [\underline{x}(\alpha) \lor \underline{y}(\alpha), \overline{x}(\alpha) \lor \overline{y}(\alpha)].$$

This operation is also visualized via Figure 1.18.

Example 1.15

Let x and y be such that

$$X = [-2,6], \quad Y = [-4,5],$$

with membership functions

$$\mu_X(x) = \begin{cases} \dfrac{x}{2} + 1, & -2 \le x \le 0, \\ -\dfrac{x}{6} + 1, & 0 \le x \le 6, \end{cases}$$

$$\mu_Y(y) = \begin{cases} \dfrac{y}{7} + \dfrac{4}{7}, & -4 \le y \le 3, \\ -\dfrac{y}{2} + \dfrac{5}{2}, & 3 \le y \le 5, \end{cases}$$

as shown in Figure 1.19.

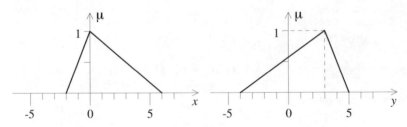

Figure 1.19 Two membership functions in Example 1.15

In the α-cut notation, following the procedure described above, one obtains

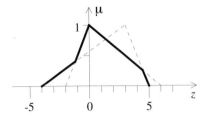

Figure 1.20 The resulting fuzzy minimum membership function

$$(X)_\alpha = [\, 2\alpha - 2, -6\alpha + 6 \,],$$

and

$$(Y)_\alpha = [\, 7\alpha - 4, -2\alpha + 5 \,],$$

so that

$$(Z)_\alpha = (X)_\alpha \wedge (Y)_\alpha$$

$$= [\, (2\alpha - 2) \wedge (7\alpha - 4), (-6\alpha + 6) \wedge (-2\alpha + 5) \,]$$

$$= \begin{cases} [7\alpha - 4, -2\alpha + 5], & 0 \le \alpha \le 1/4, \\ [7\alpha - 4, -6\alpha + 6], & 1/4 \le \alpha \le 2/5, \\ [2\alpha - 2, -6\alpha + 6], & 2/5 \le \alpha \le 1, \end{cases}$$

in which by matching $2\alpha - 2 = 7\alpha - 4$ and $-6\alpha + 6 = -2\alpha + 5$ one can find the switching points $\alpha = 1/4$ and $\alpha = 2/5$, respectively. It then yields $Z = [-4, 5]$, with

$$\mu_Z(z) = \begin{cases} -\dfrac{z}{7} + \dfrac{4}{7}, & -4 \le z \le -\dfrac{6}{5}, \\[2mm] \dfrac{z}{2} + 1, & -\dfrac{6}{5} \le z \le 0, \\[2mm] -\dfrac{z}{6} + 1, & 0 \le z \le \dfrac{9}{2}, \\[2mm] -\dfrac{z}{2} + \dfrac{5}{2}, & \dfrac{9}{2} \le z \le 5, \end{cases}$$

as shown in Figure 1.20. Graphically, if one draws horizontal lines to cut the two membership functions, then the resulting membership function has all left-hand crossing points, as shown by the solid lines in the figure.

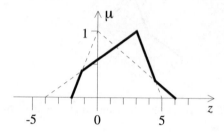

Figure 1.21 The resulting fuzzy maximum membership function

Similarly, it can be verified that

$$(Z)_\alpha = (X)_\alpha \lor (Y)_\alpha$$

$$= [\,(2\alpha - 2) \lor (7\alpha - 4),\ (-6\alpha + 6) \lor (-2\alpha + 5)\,]$$

$$= \begin{cases} [2\alpha - 2, -6\alpha + 6], & 0 \le \alpha \le 1/4, \\ [2\alpha - 2, -2\alpha + 5], & 1/4 \le \alpha \le 2/5, \\ [7\alpha - 4, -2\alpha + 5], & 2/5 \le \alpha \le 1, \end{cases}$$

and

$$\mu_Z(z) = \begin{cases} \dfrac{z}{2} + 1, & -2 \le z \le -\dfrac{6}{5}, \\[2mm] \dfrac{z}{7} + \dfrac{4}{7}, & -\dfrac{6}{5} \le z \le 3, \\[2mm] -\dfrac{z}{2} + \dfrac{5}{2}, & 3 \le z \le \dfrac{9}{2}, \\[2mm] -\dfrac{z}{6} + 1, & \dfrac{9}{2} \le z \le 6, \end{cases}$$

as shown in Figure 1.21. Graphically, if one draws horizontal lines to cut the two membership functions, then the resulting membership function has all right-hand crossing points, as shown by the solid lines in the figure.

Problems

P1.1   For a set A, the *characteristic function* of A is defined by

$$X_A(x) = \begin{cases} 1 & \text{if } x \in A, \\ 0 & \text{if } x \notin A. \end{cases}$$

For any two sets A and B and for any element $x \in S$, verify that

$$X_{A \cup B}(x) = \max\{ X_A(x), X_B(x) \},$$

$$X_{A \cap B}(x) = \min\{ X_A(x), X_B(x) \},$$

$$X_{\bar{A}}(x) = 1 - X_A(x).$$

P1.2   For $X = [1,3]$ and $Y = [3,5]$, calculate $X * Y$ for all six operations

$$* \in \{+, -, \times, \div, \max, \min\}.$$

P1.3   Consider the circuit

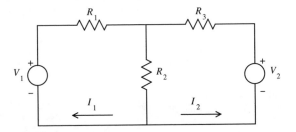

In the normal case one has

$$\begin{bmatrix} R_1 + R_2 & R_2 \\ R_2 & R_2 + R_3 \end{bmatrix} \begin{bmatrix} I_1 \\ I_2 \end{bmatrix} = \begin{bmatrix} V_1 \\ V_2 \end{bmatrix}.$$

Now, assume that $V_1 = 1$, $V_2 = 0$, but all the resistors have uncertainties, so that

$$R_1 = R_2 = R_3 = [0.8, 1.2].$$

Find the interval solution of the current vector by using the following formula (accurate to 2 decimal points):

$$\begin{bmatrix} I_1 \\ I_2 \end{bmatrix} = [R_1]^{-1} \begin{bmatrix} V_1 \\ V_2 \end{bmatrix} = \frac{\text{adj}[R_1]}{\det[R_1]} \begin{bmatrix} V_1 \\ V_2 \end{bmatrix},$$

where $\det [R_1] = (R_1 + R_2)(R_2 + R_3) - R_2^2$ and

$$R_1 = \begin{bmatrix} R_1 + R_2 & R_2 \\ R_2 & R_2 + R_3 \end{bmatrix},$$

$$\text{adj}[R_1] = \begin{bmatrix} R_2 + R_3 & -R_2 \\ -R_2 & R_1 + R_2 \end{bmatrix}.$$

P1.4 Find the best solutions for the following expressions:

(a) $Y = 1 + X + X^2 + X^3 + X^4 + X^5$ with $X = [2,3]$

(b) $Y = \dfrac{X^3 - 1}{1 - X}$ with $X = [1,5]$

P1.5 Let $X = [2,5]$ and $Y = [1,6]$ with fuzzy membership functions

$$\mu_X = \begin{cases} (x-2), & 2 \le x \le 3, \\ (5-x)/2, & 3 \le x \le 5, \end{cases}$$

and

$$\mu_Y = \begin{cases} (y-1)/3, & 1 \le y \le 4, \\ (6-y)/2, & 4 \le y \le 6, \end{cases}$$

respectively. Compute each of the following operations to obtain Z and μ_Z:

$$z = F(x,y) = x * y \quad \text{for} \quad * \in \{ +, -, \max, \min \}.$$

CHAPTER 2

Fuzzy Logic Theory

Fuzzy logic is logic. *Logic* refers to the study of methods and principles of human reasoning.

Classical logic deals with propositions (statements) that are either *true* or *false*, but not in between, nor are both simultaneous. The main content of classical logic is the study of *rules*. A rule usually takes the following form:

> Rule: IF x_1 is true AND x_2 is false AND ... AND x_n is false
> THEN y is false.

Because one, and only one, truth value (either true or false) is assumed by a logical function of a (finite) number of logical variables, the classical logic is also called a *two-valued logic*.

The fundamental assumption, upon which the classical logic is based, is that every proposition is either true or false. Yet it is now well understood and accepted that many propositions are both partially true and partially false, for which many examples can be easily given. To describe such partial truth values by some new rules, multi-valued logics were proposed and developed. At first, three-valued logic was established, which introduced a "neither" in between "true" and "false." Then, the n-valued logic of Lukasiewicz was developed (in the 1930's); when $n = \infty$, it gives the *fuzzy logic* of Zadeh (in the 1960's).

There exists a one-to-one correspondence between the two-valued logic and the classical set theory:

$$y = X_A(x) = \begin{cases} 1 & \text{if } x \in A, \\ 0 & \text{if } x \notin A, \end{cases} \Leftrightarrow y = \begin{cases} \text{true} & \text{if } x \text{ is true,} \\ \text{false} & \text{if } x \text{ is false;} \end{cases}$$

while fuzzy set and fuzzy logic has a similar correspondence.

Figure 2.1 **A console lock with a switch**

I. CLASSICAL LOGIC THEORY

A. Fundamental Concepts

Consider a typical example of a process (a system, or a plant) that is operated by a console consisting of a lock (with a key) at the high level control and an ON-OFF switch at the low level control as shown in Figure 2.1.

$$L = \begin{cases} 1 & \text{if the console is unlocked,} \\ 0 & \text{if the console is locked.} \end{cases}$$

Digits 0 and 1 are logical variables, and L, S, and P are logical functions assuming values either 0 or 1.

To establish a *truth table* (*look-up table*) for their relationships and possible outcomes of the relations, which displays all *rules* of the two-valued logic for the process, let S be the state of the switch:

$$S = \begin{cases} 1 & \text{if the switch is ON,} \\ 0 & \text{if the switch is OFF} \end{cases}$$

and let P be the status of the process:

$$P = \begin{cases} 1 & \text{if the process is running,} \\ 0 & \text{if the process is shutdown.} \end{cases}$$

Table 2.1 Truth Table for the Console-Controlled Process

Inputs			Outputs
L	S	P	P
0	0	0	0
0	0	1	1
1	0	0	0
1	0	1	0
0	1	0	0
0	1	1	1
1	1	0	1
1	1	1	1

In Table 2.1, note that if $L = 0$ (the console is locked), then the process will remain in its current status, regardless of the state of the ON-OFF switch.

A truth table is interpreted row by row. For instance, the fourth row of Table 2.1 is interpreted as follows:

> IF the console is unlocked
> AND the switch is turned to OFF
> AND the process is currently running
> THEN the process will be shutdown.

This is a *rule*. In general, a rule may have many numbers of inputs and outputs, like

> IF (input 1) AND (input 2) AND ⋯ AND (input n)
> THEN (output 1) AND (output 2) AND ⋯ AND (output m).

B. Logical Functions of the Two-Valued Logic

Primary logical functions: *negation* −, *and function* ∧, and *or function* ∨. Every other logical function is defined by composing the three primary logical

Table 2.2 Logic Functions of Two Variables

x_1	1 0 1 0	Name of Function	Symbol
x_2	1 1 0 0		
y_1	0 0 0 0	zero function	0
y_2	0 0 0 1	nor (not-or) function	$x_1 \downarrow x_2$
y_3	0 0 1 0	proper inequality	$x_1 \leftarrow x_2$
y_4	0 0 1 1	negation	$\bar{x}_2$
y_5	0 1 0 0	proper inequality	$x_1 \rightarrow x_2$
y_6	0 1 0 1	negation	$\bar{x}_1$
y_7	0 1 1 0	nonequivalence	$x_1 \not\Leftrightarrow x_2$
y_8	0 1 1 1	nand (not-and) function	$x_1 \uparrow x_2$
y_9	1 0 0 0	And function	$x_1 \wedge x_2$
y_{10}	1 0 0 1	equivalence	$x_1 \Leftrightarrow x_2$
y_{11}	1 0 1 0	identity	x_1
y_{12}	1 0 1 1	implication	$x_1 \Leftarrow x_2$
y_{13}	1 1 0 0	identity	x_2
y_{14}	1 1 0 1	implication	$x_1 \Rightarrow x_2$
y_{15}	1 1 1 0	or function	$x_1 \vee x_2$
Y_{16}	1 1 1 1	unity	1

functions $-$, $\wedge$, and $\vee$. To define a unique function, the composition order must be specified, say using parentheses. For example,

$$(\bar{x}_1 \vee \bar{x}_2) \wedge (x_1 \vee \bar{x}_3) \wedge (x_2 \vee x_3)$$

defines a unique logical function of three variables. When two logical formulas are equivalent, one usually writes, for instance,

$$(\bar{x}_1 \wedge \bar{x}_2) \vee (x_1 \wedge \bar{x}_3) \vee (x_2 \wedge x_3)$$

$$= (x_2 \wedge x_3) \vee (x_1 \wedge \bar{x}_3) \vee (\bar{x}_1 \wedge \bar{x}_2).$$

II. THE BOOLEAN ALGEBRA

A. Basic Operations of the Boolean Algebra

An algebra of the two-valued logic is the *Boolean algebra*, where only three basic logic operations are used (see Table 2.4):

Negation ¬, *and* ∧, *or* ∨.

A Boolean algebraic function can be completely described by a truth table, where all the variables are listed as inputs and the value of the logical function is the output. Conversely, Boolean algebraic functions can be written into a truth table.

For the console lock example, the second, sixth, seventh, and eighth rows of Table 2.1 can be written together in Boolean algebra as

$$P_{\text{output}} = (\bar{L} \cdot \bar{S} \cdot P) + (\bar{L} \cdot S \cdot P) + (L \cdot S \cdot \bar{P}) + (L \cdot S \cdot P),$$

which is the same as the logic formula

$$P_{\text{output}} = (\bar{L} \wedge \bar{S} \wedge P) \vee (\bar{L} \wedge S \wedge P) \vee (L \wedge S \wedge \bar{P}) \vee (L \wedge S \wedge P),$$

and they both yield the same output $P_{\text{output}} = 1$. Indeed,

$$P_{\text{output}} = (\bar{0} \cdot \bar{0} \cdot 1) + (\bar{0} \cdot 1 \cdot 1) + (1 \cdot 1 \cdot \bar{0}) + (1 \cdot 1 \cdot 1)$$

$$= (1 \cdot 1 \cdot 1) + (1 \cdot 1 \cdot 1) + (1 \cdot 1 \cdot 1) + (1 \cdot 10 \cdot 1)$$

$$= 1 + 1 + 1 + 1$$

$$= 1.$$

B. Basic Properties of the Boolean Algebra

Basic properties of the Boolean algebra are summarized in Table 2.3.

Table 2.3 Properties of the Boolean Algebra

Laws	Formulas
Characteristics	$a \cdot 0 = 0$
	$a \cdot 1 = a$
	$a + 0 = a$
	$a + 1 = 1$
Commutative Law	$a + b = b + a$
	$a \cdot b = b \cdot a$
Associative Law	$a + b + c = a + (b + c) = (a + b) + c$
	$a.b.c = a(b.c) = (a.b).c$
Distributive Law	$a.(b + c) = a.b + a.c$
Idempotence	$a.a = a$
	$a + a = a$
Negation	$\overline{\overline{a}} = a$
Inclusion	$a \cdot \overline{a} = 0$
	$a + \overline{a} = 1$
Absorptive Law	$a + a \cdot b = a$
	$a \cdot (a + b) = a$
Reflective Law	$a + \overline{a} \cdot b = a + b$
	$a \cdot (\overline{a} + b) = a.b$
	$a \cdot b + \overline{a} \cdot b \cdot c = a \cdot b + b \cdot c$
Consistency	$a \cdot b + a \cdot \overline{b} = a$
	$(a + b) \cdot (a + \overline{b}) = a$
DeMorgan's Laws	$\overline{a \cdot b} = \overline{a} + \overline{b}$
	$\overline{a + b} = \overline{a} \cdot \overline{b}$

Note that some rules of the Boolean algebra are the same as those of the ordinary algebra, such as

$$a \cdot 0 = 0, a \cdot 1 = a, \quad \text{and} \quad a + 0 = a,$$

but some are quite different, for instance,

$$a + 1 = 1.$$

Boolean algebraic laws can be used to simplify some fairly complicated logic formulas. For example:

$$P_{\text{output}} = (\overline{L} \cdot \overline{S} \cdot P) + (\overline{L} \cdot S \cdot P) + (L \cdot S \cdot \overline{P}) + (L \cdot S \cdot P)$$

$$= \overline{L} \cdot P \cdot (\overline{S} + S) + S \cdot L \cdot (\overline{P} + P)$$

[Commutative and Distributive Laws]

Table 2.4 Correspondence Among Classical Set Theory,
Boolean Algebra, and Classical Logic

Classical Set Theory	Boolean Algebra	Two-Valued Logic
∪	+	∨
∩	·	∧
−	−	−
S	1	1
∅	0	0
⊂	<	⇒
=	=	⇔

$$= \overline{L} \cdot P \cdot 1 + S \cdot L \cdot 1 \qquad \text{[Inclusion]}$$

$$= \overline{L} \cdot P + S \cdot L \qquad \text{[Characteristics]}.$$

This simplified formula is equivalent to the original one: for the second row of Table 2.1, for instance,

$$P_{\text{output}} = \overline{0} \cdot 1 + 0 \cdot 0 = 1 \cdot 1 + 0 \cdot 0 = 1 + 0 = 1,$$

and for the sixth row, one has

$$P_{\text{output}} = \overline{0} \cdot 1 + 1 \cdot 0 = 1 \cdot 1 + 1 \cdot 0 = 1 + 0 = 1.$$

Finally, note that there are interesting correspondences among the classical set theory, the Boolean algebra, and the two-valued logic, as listed in Table 2.4.

Table 2.5 The Three-Valued Logic

a	b	$\wedge$	$\vee$	$\Rightarrow$	$\Leftrightarrow$
0	0	0	0	1	1
0	1/2	0	1/2	1	1/2
0	1	0	1	1	0
1/2	0	0	1/2	1/2	1/2
1/2	1/2	1/2	1/2	1	1
1/2	1	1/2	1	1	1/2
1	0	0	1	0	0
1	1/2	1/2	1	1/2	1/2
1	1	1	1	1	1

III. MULTI-VALUED LOGIC

A. The Three-Valued Logic

In a typical theory of three-valued logic, one denotes the truth, falsity, and indeterminacy by values of 1, 0, and 1/2, respectively. The negation $\bar{a}$ of a is defined as $1 - a$. Thus,

$$\bar{1} = 0, \quad \bar{0} = 1, \quad \text{and} \quad \overline{(1/2)} = 1/2.$$

This three-valued logic is shown in Table 2.5.

B. The n-Valued Logic

For $n > 3$, an n-valued logic assumes a rational truth value in the range [0,1], defined by the following equally spaced partition:

$$0 = \frac{0}{n-1}, \frac{1}{n-1}, \frac{2}{n-1}, ..., \frac{n-2}{n-1}, \frac{n-1}{n-1} = 1.$$

Each of these truth values describes a *degree of truth*. The commonly used *n*-valued logic is attributed to Lukasiewicz-Zadeh:

$$\bar{a} \quad = \quad 1 - a,$$
$$a \wedge b \quad = \quad min\{\, a, b \,\},$$
$$a \vee b \quad = \quad max\{\, a, b \,\},$$
$$a \Rightarrow b \quad = \quad min\{\, 1, 1 + b - a \,\},$$
$$a \Leftrightarrow b \quad = \quad 1 - |\, a - b \,|.$$

When $n = \infty$, this logic does not restrict the truth values to be rational: they can be any real numbers in [0,1]. It has been shown in the logic literature that this infinite-valued logic is one-to-one corresponding to the fuzzy set theory that employs the *min, max*, and $1 - a$ operation for fuzzy set intersection, union, and complement, respectively, in the same way as the classical two-valued logic corresponds to the crisp set theory.

IV. FUZZY LOGIC AND APPROXIMATE REASONING

Fuzzy logic is logic. The ultimate goal of fuzzy logic theory is to provide a foundation for approximate reasoning using imprecise propositions based on fuzzy set theory, in a way similar to the classical reasoning using precise propositions based on the classical set theory.

To introduce this notion, it is useful to first recall how the classical reasoning works, using only precise propositions and the two-valued logic. The following syllogism is an example of such reasoning in linguistic terms:

(i)   Everyone who is 40 years old or older is old.

(ii) David is 40 years old and Mary is 39 years old.

(iii) David is old but Mary is not.

This is a very precise deductive inference, correct in the two-valued logic, though it does not make much sense in real life to have such a sharp distinction between 39 and 40.

In this classical (precise) reasoning, which uses the two-valued logic, when the (output) logical variable represented by a logical formula is always true regardless of the truth values of the (input) logical variables, it is called a *tautology*. If, on the contrary, it is always false, then it is called a *contradiction*. Various tautologies can be used to make deductive inferences, which are referred to as *inference rules*. Some of the frequently used inference rules in classical reasoning are:

modus ponens: $(a \wedge (a \Rightarrow b)) \Rightarrow b;$

modus tollens: $(\bar{b} \wedge (a \Rightarrow b)) \Rightarrow \bar{a};$

syllogism: $(a \Rightarrow b) \wedge (b \Rightarrow c) \Rightarrow (a \Rightarrow c);$

contraposition: $(a \Rightarrow b) \Rightarrow (\bar{b} \Rightarrow \bar{a}).$

The above inference rules have been commonly used in one's daily life. For example, the modus ponens is interpreted as

> IF "*a* is true" AND the statement "IF *a* is true THEN *b* is true" is true THEN "*b* is true."

Using this logic, the following deductive inference

> IF "40 years old or older is old" AND "IF 40 years old or older is old THEN David is old" THEN "David is old"

is a modus ponens, and the following deductive inference

> IF "Mary is not old" AND "IF Mary is 40 years old or older then Mary is old" THEN "Mary is not 40 years old nor older"

is a modus tollens.

Now, consider the following typical example of non-precise reasoning in linguistic terms that cannot be handled by the classical (precise) reasoning using two-valued logic:

(i) Everyone who is 40 to 70 years old is old but is very old if he (she) is 71 years old or above; everyone who is 20 to 39 years old is young but is very young if he (she) is 19 years old or below.

(ii) David is 40 years old and Mary is 39 years old.

(iii) David is old but not very old; Mary is young but not very young.

This is of course a meaningful deductive inference and has been frequently used in one's daily life. In fact, this is an example of what is called *approximate reasoning*.

Fuzzy logic can deal with such imprecise inference.

Briefly, fuzzy logic allows the imprecise linguistic terms such as

- fuzzy predicates: old, rare, severe, expensive, high, fast

- fuzzy quantifiers: many, few, usually, almost, little, much

- fuzzy truth values: very true, true, unlikely true, mostly false, false, definitely false

To describe fuzzy logic mathematically, let A be a fuzzy set - a classical set, denoted also by A, which is associated with a membership function, $\mu_A(x)$, $x \in A$. If $y = \mu_A(x_0)$ is a point in [0,1], representing the truth value of the proposition "x_0 is a," or simply "a," then the truth value of "not a" is given by

$$\bar{y} = \mu_A(x_0 \text{ is not } a) = 1 - \mu_A(x_0 \text{ is } a) = 1 - \mu_A(x_0) = 1 - y.$$

Consequently, for n members x_1, ..., x_n in A with n corresponding truth values, $y_i = \mu_A(x_i)$ in [0,1], $i = 1,...,n$, the *truth values* of "not a" is defined as

$$\bar{y}_i = 1 - y_i, \quad i = 1,...,n.$$

Note that actually $n = \infty$ is allowed. With $n > 3$, the logical operations *and, or, not, implication*, and *equivalence* are defined as follows: for any $a, b \in A$,

$$\mu_A(a \wedge b) := min\{ \mu_A(a), \mu_A(b) \} = \mu_A(a) \wedge \mu_A(b);$$
$$\mu_A(a \vee b) := max\{ \mu_A(a), \mu_A(b) \} = \mu_A(a) \vee \mu_A(b);$$
$$\mu_A(\bar{a}) := 1 - \mu_A(a);$$
$$\mu_A(a \Rightarrow b) := min\{ 1, 1 + \mu_A(b) - \mu_A(a) \} = \mu_A(a) \Rightarrow \mu_A(b);$$
$$\mu_A(a \Leftrightarrow b) := 1 - |\mu_A(a) - \mu_A(b)| = \mu_A(a) \Leftrightarrow \mu_A(b).$$

For multi-point cases, for instance, $a_i, b_j \in A$, with $\mu_A(a_i), \mu_A(b_j) \in [0,1]$, $i = 1,...,n$, $j = 1,...,m$, where $1 \le n, m \le \infty$, the following "max-min" formula is used for two fuzzy sets: $a \wedge b$:

$$\mu_A(a_1,...,a_n; b_1,...,b_m) = \mu_A(a_1,...,a_n) \wedge \mu_A(b_1,...,b_m),$$

where $a := \{a_1,...,a_n\}$ and $b := \{b_1,...,b_m\}$.

Other operations are defined accordingly.

In the classical two-valued logic, for instance in the modus ponens, the inference rule is

$$(a \wedge (a \Rightarrow b)) \Rightarrow b.$$

In terms of membership values, this is equivalent to

IF $\mu(a) = 1$
   AND $\mu(a{\Rightarrow}b) = min\{\ 1,\ 1 + \mu(b) - \mu(a)\ \} = min\{\ 1,\ \mu(b)\} = 1$
THEN $\mu(b) = 1$.

Since if $\mu(b) \neq 1$, then it will contradict either $\mu(a) = 1$ or $\mu(a{\Rightarrow}b) = 1$.

In fuzzy logic, the inference rule reads the same: for the modus ponens, it is

$$(a \wedge (a \Rightarrow b)) \Rightarrow b.$$

But, in terms of membership values, one has

IF $\mu(a) > 0$ AND $\mu(a{\Rightarrow}b) = min\{\ 1,\ 1 + \mu(b) - \mu(a)\ \} > 0$
THEN $\mu(b) > 0$.

Here, $\mu > 0$ means $\mu \in (0,1]$. This is a *fuzzy modus ponens*.

The above fuzzy logic inference can be interpreted as follows:

IF a is true with a certain degree of confidence AND "IF a is true with a certain degree of confidence THEN b is true with a certain degree of confidence" THEN b is true with a certain degree of confidence.

As will be seen below, all these "degrees of confidence" can be quantitatively evaluated by using the corresponding membership functions.

Note that in the extreme cases, fuzzy logic is consistent with the classical logic. This can be easily verified by replacing " > 0" with " = 1" in the above fuzzy logic inference.

Example 2.1

In the classical two-valued logic, the modus tollens is

$$(\bar{b} \wedge (a \Rightarrow b)) \Rightarrow \bar{a}.$$

The following is a simple case:

Premise	David cannot work
Implication	If David is young then he can work
Conclusion	David is not young

This example does not make much sense, of course, since it implies that if David is not young then he cannot work at all. Thus, one sees a limitation of the two-valued logic in real-life applications. The problem lies in that two-valued logic can only describe "young or old," "can or cannot," etc. Indeed, this is the best one can do for this example using two-valued logic.

In contrast, the fuzzy modus tollens has the same rule:

$$(\overline{b} \wedge (a \Rightarrow b)) \Rightarrow \overline{a} \, ,$$

but it provides a much more meaningful inference as follows, which only implies that David cannot work so hard:

Premise	David cannot work much
Implication	If David is much younger then he can work more
Conclusion	David is not so young

For examples like this, one only needs to select reasonable membership functions to describe "young, very young, old, very old, much, much more, hard, very hard," etc., such that they are meaningful and practical for the applications in consideration. After that, one can follow rigorous mathematical and logical formulas and rules to carry out all subsequent calculations and inferences, thereby reaching a correct and meaningful answer to the question at hand.

V. FUZZY RELATIONS

Let A and B be fuzzy sets. A *fuzzy relation* is a relation between elements of A and elements of B, described by a membership function, $\mu_{A \times B}(a,b)$, where $a \in A$ and $b \in B$.

A simple example of a fuzzy relation is the following:

Let $A = B = [0,1] \subset \mathrm{R}$. Define a membership function $\mu_{A \times B}(a,b)$ for the relation "a is slightly larger than b" by (see Figure 2.2)

$$\mu_{A \times B}(a,b) = \begin{cases} 0 & a \leq b, \\ e^{-(a-b)} & a > b. \end{cases}$$

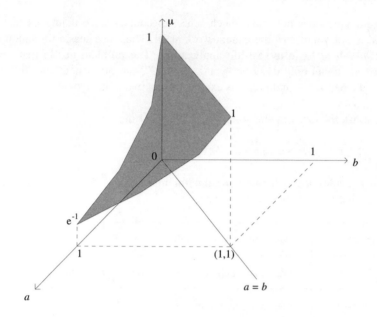

Figure 2.2 The fuzzy relation "*a* is slightly larger than *b*"

Then, A, B, and $\mu_{A \times B}$ together define a fuzzy relation between $a \in A$ and $b \in B$.

For discrete supports, fuzzy relations can be defined by tables or matrices. For example, let $A = \{\, a_1, a_2, a_3, a_4 \,\} = \{\, 1, 2, 3, 4 \,\}$ and $B = \{\, b_1, b_2, b_3 \} = \{\, 0,\ 0.1,\ 2\ \}$. Then, the following table defines a fuzzy relation "*a* is considerably larger than *b*." This table is a *discrete membership function*, with two variables a and b and 12 membership values given inside the table.

	b_1	b_2	b_3
a_1	0.6	0.6	0.0
a_2	0.8	0.7	0.0
a_3	0.9	0.8	0.4
a_4	1.0	0.9	0.5

Example 2.2

Let $A, B \subseteq R$. Consider two fuzzy subsets:

   (i) A with an associate membership function, $\mu_A(a)$, $a \in A$;

  (ii) B with an associate membership function, $\mu_B(b)$, $b \in B$.

Let $\mu_{A \times B}(a,b)$ be a membership function defined on $A \times B$ by

$$\mu_{A \times B}(a,b) = min\{ \mu_A(a), \mu_B(b) \}, \quad (a,b) \in A \times B.$$

Then, A, B, and $\mu_{A \times B}$ together define a fuzzy relation.

For notational convenience, denote by $R_{(A,B)}$ a fuzzy relation defined on $A \times B$ where A and B are fuzzy sets associated with a membership function, $\mu_{A \times B}$.

For two fuzzy relations, $R^1_{(A,B)}$ and $R^2_{(A,B)}$, defined on $A \times B$, their *union* and *intersection*, $R^1_{(A,B)} \cup R^2_{(A,B)}$ and $R^1_{(A,B)} \cap R^2_{(A,B)}$, are defined, respectively, by the membership functions

$$\mu_{R^1 \cup R^2}(a,b) = max\{ \mu_{R^1}(a,b), \mu_{R^2}(a,b) \}, \quad (a,b) \in A \times B;$$

$$\mu_{R^1 \cap R^2}(a,b) = min\{ \mu_{R^1}(a,b), \mu_{R^2}(a,b) \}, \quad (a,b) \in A \times B,$$

Example 2.3

Let

$$A = \{ a_1, a_2, a_3, a_4 \} = \{ 1, 2, 3, 4 \}$$

and

$$B = \{ b_1, b_2, b_3 \} = \{ 0, 0.1, 2 \}.$$

Let $R^1_{(A,B)}$ be the fuzzy relation "a is considerably larger than b" defined by the following table (discrete membership function):

	b_1	b_2	b_3
A_1	0.6	0.6	0.0
A_2	0.8	0.7	0.0
A_3	0.9	0.8	0.4
A_4	1.0	0.9	0.5

$R^1_{(A,B)}$:

Let $R^2_{(A,B)}$ be the fuzzy relation "a is considerably close to b" defined by the following table (discrete membership function):

	b_1	b_2	b_3
a_1	0.2	0.2	0.5
a_2	0.1	0.1	1.0
a_3	0.0	0.0	0.5
a_4	0.0	0.0	0.3

$R^2_{(A,B)}$:

Then, $R^1_{(A,B)} \cup R^2_{(A,B)}$ and $R^1_{(A,B)} \cap R^2_{(A,B)}$ are given by the following two tables, respectively:

	b_1	b_2	b_3
a_1	0.6	0.6	0.5
a_2	0.8	0.7	1.0
a_3	0.9	0.8	0.5
a_4	1.0	0.9	0.5

$R^1_{(A,B)} \cup R^2_{(A,B)}$:

	b_1	b_2	b_3
a_1	0.2	0.2	0.0
a_2	0.1	0.1	0.0
a_3	0.0	0.0	0.4
a_4	0.0	0.0	0.5

$R^1_{(A,B)} \cap R^2_{(A,B)}$:

Here, in this example, $R^1_{(A,B)} \cup R^2_{(A,B)}$ means that "a is either considerably larger than or considerably close to b," whereas $R^1_{(A,B)} \cap R^2_{(A,B)}$ means that "a is considerably larger than, as well as considerably close to b." The latter statement is somewhat self-contradictory but is acceptable by fuzzy relations, which, however, has small membership values in general.

Let $R^1_{(A,B)}$ and $R^2_{(B,C)}$ be two fuzzy relations. Their *max-min* composition is defined to be the new relation

$$R_{(A,B,C)} = R^1_{(A,B)} \circ R^2_{(B,C)}$$

where $\circ$ indicates the composition of two relations, with the membership function defined and computed by

$$\mu_{R^1 \circ R^2}(a,c) = \max_{b \in B} \{ \min\{ \mu_{R^1}(a,b), \mu_{R^2}(b,c) \} \}, \quad (a,c) \in A \times C.$$

The following example shows an application of the fuzzy relations and max-min compositions for approximate reasoning.

Example 2.4

Let $A = B = \{1, 2, 3, 4\}$, with the following discrete membership function for the fuzzy description "small":

$$\mu_A(a) = \begin{cases} 1.0 & \text{if } a=1, \\ 0.7 & \text{if } a=2, \\ 0.3 & \text{if } a=3, \\ 0.0 & \text{if } a=4. \end{cases}$$

Let R be a fuzzy relation between two members in A, meaning "approximately equal," and be described by the following table (discrete membership function):

		1	2	3	4
	1	1.0	0.5	0.0	0.0
$R:$	2	0.5	1.0	0.5	0.0
	3	0.0	0.5	1.0	0.5
	4	0.0	0.0	0.5	1.0

Suppose that one wants to perform the following fuzzy logic inference (approximate reasoning):

Premise	a is small
Implication	a and b are approximately equal
Conclusion	b is somewhat small

Then, one can apply the *max-min* composition of the fuzzy relations as follows:

(i)   "a is small": $\mu_A(a)$ is available;

(ii) "a and b are approximately equal": $\mu_R(a,b)$ is given by the table;

(iii) let $\mu_B(b)$ be the membership function for the conclusion (a fuzzy modus ponens for this example):

$$\mu_B(b) = \max_{a \in A} \{ \min\{ \mu_A(a),\ \mu_R(a,b) \} \}, \quad b \in B = A.$$

For instance, for $b = 2$, the result is

$$\mu_B(2) = \max_{a \in A} \{ \min\{ \mu_A(a),\ \mu_R(a,2) \} \}$$

$$= \max\{ \min\{1.0,0.5\},\ \min\{0.7,1.0\},\ \min\{0.3,0.5\},\ \min\{0.0,0.0\}\}$$

$$= \max\{ 0.5,\ 0.7,\ 0.3,\ 0.0 \} = 0.7.$$

Similarly, one can evaluate $\mu_B(1)$, $\mu_B(3)$, and $\mu_B(4)$. The final result is

$$\mu_B(b) = \begin{cases} 1.0 & \text{if } b=1, \\ 0.7 & \text{if } b=2, \\ 0.5 & \text{if } b=3, \\ 0.3 & \text{if } b=4. \end{cases}$$

From the above simple and illustrative examples, one can see that fuzzy logic does work for approximate reasoning within a mathematical framework, so that even computer programs can be written to implement it. This is important and useful for computer decision-making and artificial intelligence. Some practical examples of fuzzy logic in these kinds of applications will be further discussed in the next chapter.

Problems

P2.1   Compute the following logical operation:

"IF $\mu(a) = 0.9$ AND $\mu(a \Rightarrow b) = 0.3$ THEN $\mu(b) = $?"

P2.2   In Example 2.4, it has been shown how to compute the membership value $\mu_B(b = 2) = 0.7$. Verify if the rest results given in the example are correct:

$$\mu_B(b = 1) = 1, \ \mu_B(b = 3) = 0.5, \ \mu_B(b = 4) = 0.3.$$

P2.3   Let $A = B = \{\ 1, 5, 9, 10\ \}$ be some typical job performance indexes in an application, with the following discrete membership function for the fuzzy description "poor performance":

$$\mu_A(a) = \begin{cases} 1.0 & \text{if } a=1, \\ 0.5 & \text{if } a=5, \\ 0.1 & \text{if } a=9, \\ 0.0 & \text{if } a=10. \end{cases}$$

Let R be a fuzzy relation between two members in A, meaning "very close to each other," and be defined by the following table:

		1	5	9	10
	1	1.0	0.5	0.0	0.0
R:	5	0.5	1.0	0.5	0.1
	9	0.0	0.5	1.0	0.5
	10	0.0	0.1	0.5	1.0

Suppose that one wants to perform the following fuzzy logic inference (approximate reasoning):

Premise	a has poor performance
Implication	a and b are very close to each other
Conclusion	b has somewhat poor performance

Compute its membership value $\mu_B(5) = ?$

P2.4 Fill in the blanks ...... in the following table, given the modus tollens:

$$(\overline{b} \wedge (a \Rightarrow b)) \Rightarrow \overline{a}$$

where a means "the price is cheap" and b means "the car is affordable"

Premise	The car is ......
Implication	If ...... then ......
Conclusion	The price is ......

P2.5 Calculate the confidence values of the following implication statements:

$$\text{IF } \mu(\overline{a}) = 0.7 \text{ AND } \mu(a \Rightarrow b) = 0.8 \text{ THEN } \mu(b) = ?$$

P2.6 Are the following statements meaningful, and why?

(a) IF I can guess today's weather is good in London with confidence value 0.9 THEN I am sure I can predict that tomorrow's weather is also good in Hong Kong with confidence value 0.1.

(b) IF I can guess today's weather is good in London with confidence value 0.1 THEN I am sure I can predict that tomorrow's weather is also good in Hong Kong with confidence value 0.9.

P2.7 Write a Rule Base for the following fuzzy logic implication:

$$a \text{ is } A \text{ UNLESS } b \text{ is } B \text{ OR } c \text{ is } C$$

CHAPTER 3

Some Applications of Fuzzy Logic

The main intention of using fuzzy logic in industrial applications is to provide some efficient methodologies for computer-based diagnosis, information processing, decision-making, as well as approximate reasoning in designing practical computational schemes using imprecise criteria and inaccurate data to solve some real-world technological problems. Such methodologies use fuzzy set theory and fuzzy logic principles, in a way similar to human experts' reasoning and decision-making, thereby achieving some kind of artificial intelligence.

A few application examples of fuzzy logic are introduced in this chapter, so as to provide some flavor of the flexibility and wide-range applicability of fuzzy logic in industrial, financial, and management practice.

I. PRODUCT QUALITY EVALUATION

Product quality evaluation is an example where fuzzy logic finds its applications. Quality evaluation is usually the last but also the most important stage in an industrial production process. Since there are always some deviations in the quality of industrial products, particularly new products of manufacturing, quality evaluation and control is an essential step in the production process. This section discusses a simplified version of fuzzy-logic-based quality evaluation example.

Although there are many causes of deviations in product quality, they are generally categorized into two main types: chance cause and assignment cause. Here, assignable causes can be handled with a good design of quality evaluation and control procedure, subject to some reasonable pre-set criteria, and oftentimes can be made automatic via computer programming.

Today, most quality evaluation processes use linguistic terms, such as "very good" and "quite poor." Whether or not a product has "good quality" depends on a certain criteria set by a human expert. For quality evaluation, one usually sets some criteria for the measurement of various aspects of the quality of a product, as shown in Figure 3.1, where the membership functions indicate if a sampled product is acceptable or not by the pre-set standards. These membership functions describe the features and sizes of the product and are given by

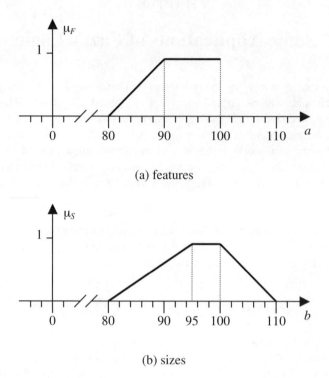

(a) features

(b) sizes

Figure 3.1 Membership functions for product quality evaluation

$$\mu_F(a) = \begin{cases} 0 & 0 \le a \le 80 \\ (a-80)/10 & 80 \le a \le 90 \\ 1 & 90 \le a \le 100 \end{cases}$$

$$\mu_S(b) = \begin{cases} 0 & 0 \le b \le 80 \\ (b-80)/15 & 80 \le b \le 95 \\ 1 & 95 \le b \le 100 \\ (110-b)/10 & 100 \le b \le 110 \end{cases}$$

For example, "features" of a product can be its designed colors, shape, weight, and material used; and "sizes" of the product can be its x, y, z dimensions.

To see how the above quality membership functions can be used for quality evaluation and control, consider a test on a set of a product consisting of 5 samples, with scores of "features" and "X-Y-Z sizes" against the pre-set standards as shown in Table 3.1. Let the sampled product set be

Table 3.1 Features and sizes of the sampled products

	Features (a)	X size (b_x)	Y size (b_y)	Z size (b_z)
P_1	86	97	95	102
P_2	98	89	98	90
P_3	90	92	96	88
P_4	96	96	102	104
P_5	90	103	100	101

Table 3.2 Membership values of features and sizes of the sampled products

	Features $\mu_F(a)$	X size $\mu_S(b_x)$	Y size $\mu_S(b_y)$	Z size $\mu_S(b_z)$
P_1	0.6	1.0	1.0	0.80
P_2	1.0	0.6	1.0	0.67
P_3	1.0	0.8	1.0	0.53
P_4	1.0	1.0	0.8	0.60
P_5	1.0	0.7	1.0	0.90

$$A = \{ P_1, P_2, P_3, P_4, P_5 \}.$$

According to the two membership functions given above, the corresponding membership values are obtained from the figures as shown in Table 3.2.

If all columns in Table 3.2 are considered equally important in one's decision, then a simple and naïve approach is to calculate each individual's total membership values and then compare them. This gives

$$P_1 = 0.6 + 1.0 + 1.0 + 0.80 = 3.40$$

$$P_2 = 1.0 + 0.6 + 1.0 + 0.67 = 3.27$$

$$P_3 = 1.0 + 0.8 + 1.0 + 0.53 = 3.33$$

$$P_4 = 1.0 + 1.0 + 0.8 + 0.60 = 3.40$$

$$P_5 = 1.0 + 0.7 + 1.0 + 0.90 = 3.60$$

In conclusion, product sample x_5 has the highest score overall, and therefore is the best among the five samples.

It should be noted that when the number of samples is large, this kind of quality evaluation cannot be handled manually, hence computers are needed. The above analysis and computation procedure can be written as a program on a computer, so that any set of product samples can be evaluated by the program, to check which sampled piece of the product has the best quality. This can save a lot of time and labor for human operators in many kinds of time-consuming, routine, and tedious work like this.

II. DECISION-MAKING FOR INVESTMENT

A typical example of this kind is a simplified model of e-trade software design for stocks.

Suppose that an investor wants to write a simple PC program that can make a suggestion to him each day whether he should buy a certain stock, on the basis of the available daily information data about that particular stock in the market.

One possible approach is as follows. Suppose that the investor has evaluated sufficient historical data and obtained the limits for stock values within the interval $A = [0, 13]$, with the following membership function for the fuzzy description "stock index is low":

$$\mu_A(a) = \begin{cases} 1.0 & \text{if} & 0 \le a \le 10, \\ 0.7 & \text{if} & 10 < a \le 12, \\ 0.3 & \text{if} & 12 < a \le 13, \\ 0.0 & \text{if} & 13 < a. \end{cases}$$

Let R be a fuzzy relation "approximately equal" between a member $a \in A$ and a member of the pre-set list of reference indexes { 9, 10, 11, 12 } (rounded for simplicity here), with a membership function defined by Table 3.3:

Table 3.3 A discrete membership function for "approximately equal"

	$0 \leq a \leq 10$	$10 < a \leq 12$	$12 < a \leq 13$	$13 < a$
9	1.0	0.8	0.3	0.0
10	1.0	0.9	0.7	0.3
11	0.8	1.0	0.9	0.7
12	0.5	1.0	1.0	0.9

Suppose that the investor wants to perform the following fuzzy logic inference (approximate reasoning) using the program and let the PC make a suggestion for him for today's action (*buy* or *not buy* the stock) based on today's available market data $b \in \{ 9, 10, 11, 12 \}$:

Premise	*a* is low
Implication	*a* and *b* are approximately equal
Conclusion	*b* is somewhat low

In addition, of course, the investor needs to set a rule for the program to make a suggestion. This *rule* can be, for instance:

"IF *b* is low with confidence value 0.9 or above THEN buy the stock."

On the PC, the designed computing program applies the *max-min* composition on the fuzzy relations (see Section V, Chapter 2):

(i) "*a* is low": $\mu_A(a)$ is available;

(ii) "*a* and *b* are approximately equal": $\mu_R(a,b)$ is given by the table;

(iii) Let $\mu_B(b)$ be the membership function for the conclusion:

$$\mu_B(b) = \max_{a \in A} \{ \ min\{ \ \mu_A(a), \ \mu_R(a,b) \ \} \ \} \ .$$

If, say, $b = 11$, then the result is:

$$\mu_B(11) \;=\; \max_{a \in A} \{\; min\{\; \mu_A(a),\; \mu_R(a,11) \;\} \;\}$$

$$= max\{\; min\{1.0,0.8\},\, min\{0.7,1.0\},\, min\{0.3,0.9\},\, min\{0.0,0.7\}\}$$

$$= max\{\; 0.8,\, 0.7,\, 0.3,\, 0.0 \;\}$$

$$= 0.8.$$

Therefore, the program will make the following suggestion:

> "The confidence 0.8 is below the threshold. Do not buy the
> stock today."

This conclusion really depends on how the investor pre-set the threshold (preference) and how to select a rule for decision-making. This decision is meaningful only if the above pre-set preference and selected decision rule are acceptable.

This simplified example shows that the final decision is ultimately made by the user, whereas the computer simply provides assistance in relaxing human from some routine, tedious, and huge amount of work.

III. PERFORMANCE EVALUATION

An example of a company's business performance evaluation is now discussed, showing how some qualitative comparisons may be quantified, if needed.

It is clear that there is no precise and sharp criterion to measure a company's business performance in an absolutely crispy manner. One usually says so-and-so is "excellent" or "very good" or "unsatisfactory." But it is difficult to come out with a quantified measurement to tell why. Also difficult is how to compare two companies to determine which one is better, based on their performance measured in some terms that can be quantified. Similar problems exist in a company's evaluation of its employees' performance. In such a difficult and linguistically described task of performance evaluation, fuzzy logic can be helpful.

A. Problem Formulation

Suppose that the three fuzzy membership functions shown in Figure 3.2 describe the business performance of a company, regarding its service performance (from customers' feedback), cooperation with other companies, and sale profits:

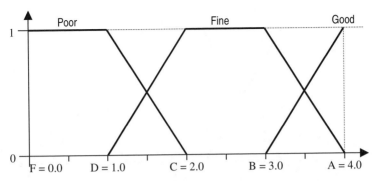

Figure 3.2 (a) Membership functions for service performance

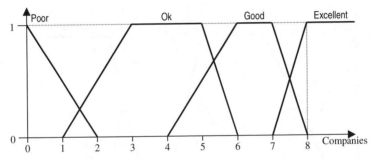

Figure 3.2 (b) Membership functions for inter-company's cooperation

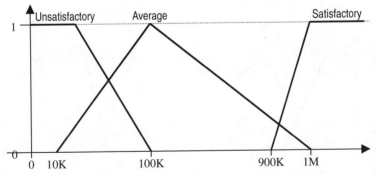

Figure 3.2 (c) Membership functions for sale profits

As can be seen, the three figures have different dimensions and units. It is convenient to first normalize them into a unified and comparable numerical scale, with the same size for comparability and also for convenience of calculation. This can be easily done, yielding three normalized membership functions as shown in Figure 3.3, which are denoted by $\mu_S(\cdot)$, $\mu_C(\cdot)$, and $\mu_P(\cdot)$, respectively.

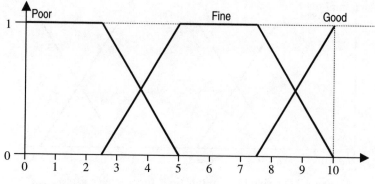

(a) Normalized membership functions for service performance

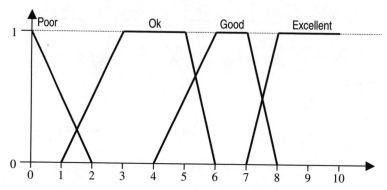

(b) Normalized membership functions for inter-company's cooperation

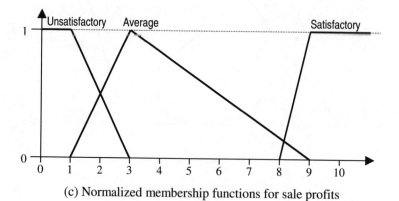

(c) Normalized membership functions for sale profits

Figure 3.3. Various normalized membership functions.

B. Performance Evaluation Formula

The overall *performance index* of the company is calculated by

$$P_{index} = \frac{\mu_S(x_S) \cdot x_S + \mu_C(x_C) \cdot x_C + \mu_P(x_P) \cdot x_P}{\mu_S(x_S) + \mu_C(x_C) + \mu_P(x_P)}$$

where $x_S \in X_S = [0,10]$, $x_C \in X_C = [0,10]$, $x_P \in X_P = [0,10]$, and $\mu_T(x_T)$ etc. are their corresponding membership values.

This is the commonly used "weighted average" formula for the final output (performance index) derived from the three inputs (service performance, inter-company cooperation, and sale profits). This weighted average formula can be rewritten as

$$P_{index} = w_S \cdot x_S + w_C \cdot x_C + w_P \cdot x_P$$

with

$$w_S = \frac{\mu_S(x_S)}{\mu_S(x_S) + \mu_C(x_C) + \mu_P(x_P)},$$

$$w_C = \frac{\mu_C(x_C)}{\mu_S(x_S) + \mu_C(x_C) + \mu_P(x_P)},$$

$$w_P = \frac{\mu_P(x_P)}{\mu_S(x_S) + \mu_C(x_C) + \mu_P(x_P)},$$

$$w_S + w_C + w_P = 1.$$

Hence, it is a convex combination of the three factors, which determines the effects of the three factors according to the values of their respective membership functions.

It is important to note that if a point, say x_S, has two different membership values, then both of the two values should be counted in the above weighted average formula, namely, this point will be evaluated twice in the formula (treated as two independent situations).

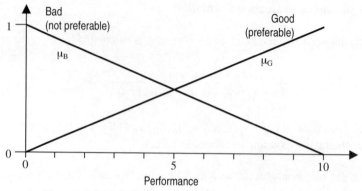

Figure 3.4 (a) Membership function for "performance"

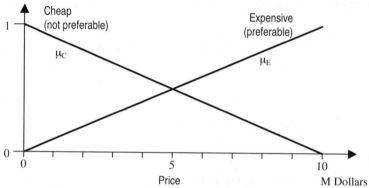

Figure 3.4 (b) Membership function for the "price"

The above evaluation yields a unique and well-balanced performance index for a company, so that the performance of each company can be measured and then compared. Well-performed companies can then be rewarded according to their index values as compared to other companies or to a pre-set threshold.

In the above evaluation, two companies with very close inputs usually yield very close outputs. However, it is quite possible that sometimes one company with slightly better inputs yields a slightly worse output (final performance evaluation result). This is because the given information is somewhat inaccurate and vague, namely, the "slightly better inputs" may not actually be slightly better, and the membership functions used may not be perfectly reasonable. Nevertheless, if this indeed happens, the corresponding confidence value will be very low, meaning that the incorrect conclusion is not very reliable in general. This is similar to some common human errors: in such a practically indistinguishable situation, even individuals can make a small mistake in judgment. Generally, these kinds of seemingly small errors are unavoidable in fuzzy-logic-based as well as other types of artificial intelligence.

Finally, it must be pointed out that a common technical mistake in such performance evaluation is illustrated as follows.

Suppose that one has two criteria for evaluating a product: "performance" and "price."

Let the "performance" measurement be as shown in Figure 3.4 (a), where a small measurement number such as 1 and 2 means "bad" while a large measurement number such as 9 and 10 means "good." As usual, two fuzzy membership functions are chosen: μ_B is the "bad" membership function and μ_G is the "good" membership function. Here, a large measurement number is preferable but a small one is not.

Let the "price" measurement be as shown in Figure 3.4 (b), where a small price number such as 1 and 2 means "cheap" while a large price number such as 9 and 10 means "expensive." Similarly, two fuzzy membership functions are chosen: μ_C is the "cheap" membership function and μ_E is the "expensive" membership function. Here, a small price number is preferable but a large one is not – exactly *opposite* to the "performance" measurement above.

Now, suppose that someone has a product with "performance = 9," which is preferable, and "price = 8" and this is not preferable.

When coming to the final evaluation, one may use the following formula and get

$$\text{Output} \;=\; \frac{9 \times \mu_G(9) + 9 \times \mu_B(9) + 8 \times \mu_E(8) + 8 \times \mu_C(8)}{\mu_G(9) + \mu_B(9) + \mu_E(8) + \mu_C(8)}$$

$$=\; \frac{9 \times 0.9 + 9 \times 0.1 + 8 \times 0.8 + 8 \times 0.2}{0.9 + 0.1 + 0.8 + 0.2}$$

$$=\; 8.5$$

However, 8×0.8 here is a large number contributing considerably to the final result 8.5, which is the decision index. Actually, this number means "expensive" and is not preferable, so it should not contribute so much to the final index for decision-making. For instance, one may compare it to the case of "performance = 9" and "price = 1," which yields

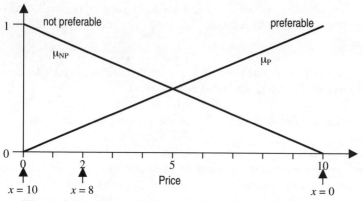

Figure 3.4 (c) The re-defined membership function for "price"

$$\text{Output} = \frac{9 \times \mu_G(9) + 9 \times \mu_B(9) + 1 \times \mu_E(1) + 1 \times \mu_C(1)}{\mu_G(9) + \mu_B(9) + \mu_E(1) + \mu_C(1)}$$

$$= \frac{9 \times 0.9 + 9 \times 0.1 + 1 \times 0.1 + 1 \times 0.9}{0.9 + 0.1 + 0.1 + 0.9}$$

$$= 5$$

In this comparison, one might conclude that the first piece of product is better than the second one (8.5 > 5). But clearly this is ridiculous, since the second piece of product has the same good performance but with a lower price!

The problem here lies in that the first preferable membership function and second non-preferable membership function are both located on the right-hand side in Figures 3.4 (a) and (b), respectively. Something is wrong here.

To fix the problem, one should keep the first figure unchanged, but reverse the second figure in such a way that the two preferable membership functions are coherent. There are several different ways to do so; one way is to re-define the second membership function as shown in Figure 3.4 (c):

In Figure 3.4 (c), one has changed the name of the membership functions to be μ_P for "preferable" and μ_{NP} for "not preferable" in order to avoid possible confusion, and, correspondingly, converted x to $10 - x$, namely, $10 \rightarrow 0$, $9 \rightarrow 1$, ..., $0 \rightarrow 10$. Consequently, the first piece of product yields

$$\text{Output} = \frac{9 \times \mu_G(9) + 9 \times \mu_B(9) + 2 \times \mu_{NP}(2) + 2 \times \mu_P(2)}{\mu_G(9) + \mu_B(9) + \mu_{NP}(2) + \mu_P(2)}$$

$$= \frac{9 \times 0.9 + 9 \times 0.1 + 2 \times 0.8 + 2 \times 0.2}{0.9 + 0.1 + 0.8 + 0.2}$$

$$= 5.5$$

while the second one yields

$$\text{Output} = \frac{9 \times \mu_G(9) + 9 \times \mu_B(9) + 9 \times \mu_{NP}(9) + 9 \times \mu_P(9)}{\mu_G(9) + \mu_B(9) + \mu_{NP}(9) + \mu_P(9)}$$

$$= \frac{9 \times 0.9 + 9 \times 0.1 + 9 \times 0.1 + 9 \times 0.9}{0.9 + 0.1 + 0.1 + 0.9}$$

$$= 9.0$$

As a result, one concludes that the second piece of the product is better, which is a correct answer.

The above simple example provides a basic principle for the general case in a practical design: One must use technical knowledge and common sense to determine a reasonable working approach, especially with a correct average formula, for the application at hand. Once again, this is a *design* problem for which there does not exist a universal solution and the designer must use common sense and scientific knowledge to generate a good proposal that works for the intended application.

IV. MISCELLANEOUS EXAMPLES

This section provides a few more examples of fuzzy-logic-based decision-making for practical applications.

Example 3.1 Network Traffic Control

To minimize the delay and traffic congestion in a computer network, it is preferable to choose a route that has minimum delay and bandwidth utilization. For this purpose, one needs an index that can describe the traffic condition at each node during the route setup.

Performance evaluation by using fuzzy logic may be employed to find the desired index that can describe the traffic condition at a node. The traffic condition depends on delay and bandwidth utilization, which in turn depend on the buffer queue length and the traffic type. It is a somewhat complicated task to describe the traffic condition by directly using these parameters in a

Input rate: Unlimited
Buffer size: 200 cells
Output rate: 1 cell per time slot

Figure 3.5 1-input and 1-output FIFO node model

traditional manner. However, establishing a performance index using fuzzy logic may provide a feasible solution to the problem. In this simplified example, fuzzy logic is employed to find the index that describes the traffic condition according to the parameters of the buffer queue length, bandwidth utilization, and traffic type at a node.

For simplicity, a 1-input and 1-output FIFO node model is used (Figure 3.5). In this example:

Buffer Queue Length

A longer queue length means a longer delay and it is more likely to have data loss at a burst increase of input traffic. Hence, a short queue is desirable.

An average queue length provides information about the average delay at the node, while a burst queue length indicates the momentary delay at the node. As both the average and the burst queue lengths give information about the traffic condition, they both will be considered in this study.

Figure 3.6 (a)-(b) are the chosen membership functions for the average queue length and the burst queue length, respectively.

After normalization, one has the membership function for the average queue length as

$$\mu_{QA,e}(x) = \begin{cases} 1 & 0 \leq x \leq 2 \\ 2 - \dfrac{x}{2} & 2 \leq x \leq 4 \\ 0 & 4 \leq x \leq 10 \end{cases}$$

$$\mu_{QA,f}(x) = \begin{cases} 0 & 0 \leq x \leq 6 \\ \dfrac{x}{2} - 3 & 6 \leq x \leq 8 \\ 1 & 8 \leq x \leq 10 \end{cases}$$

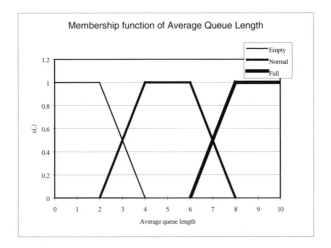

Figure 3.6 (a) The membership function for the average queue length

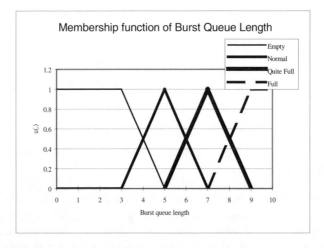

Figure 3.6 (b) The membership function for the burst queue length

$$
\mu_{QA,n}(x) = \begin{cases}
0 & 0 \le x \le 2 \\
\dfrac{x}{2} - 1 & 2 \le x \le 4 \\
1 & 4 \le x \le 6 \\
4 - \dfrac{x}{2} & 6 \le x \le 8 \\
0 & 8 \le x \le 10
\end{cases}
$$

and the membership function for   the burst queue length as

$$\mu_{QB,e}(x) = \begin{cases} 1 & 0 \le x \le 3 \\ 2.5 - \dfrac{x}{2} & 3 \le x \le 5 \\ 0 & 5 \le x \le 10 \end{cases}$$

$$\mu_{QB,qf}(x) = \begin{cases} 0 & 0 \le x \le 5 \\ \dfrac{x}{2} - 2.5 & 5 \le x \le 7 \\ 4.5 - \dfrac{x}{2} & 7 \le x \le 9 \\ 0 & 9 \le x \le 10 \end{cases}$$

$$\mu_{QB,n}(x) = \begin{cases} 0 & 0 \le x \le 3 \\ \dfrac{x}{2} - 1.5 & 3 \le x \le 5 \\ 3.5 - \dfrac{x}{2} & 5 \le x \le 7 \\ 0 & 7 \le x \le 10 \end{cases}$$

$$\mu_{QB,f}(x) = \begin{cases} 0 & 0 \le x \le 7 \\ \dfrac{x}{2} - 3.5 & 7 \le x \le 9 \\ 1 & 9 \le x \le 10 \end{cases}$$

Bandwidth Utilization

Bandwidth utilization shows the crowding condition at the node. Higher bandwidth utilization means higher risk for a congestion caused by a momentary increase of input traffic. Hence, low utilization is desirable if delay time is considered. Higher utilization is desirable if link efficiency or lost effectiveness is considered. However, very high utilization may lead to data loss as the buffer is overflow.

Figure 3.7 (a)-(b) are the chosen membership functions for the bandwidth utilization:

After normalization, one obtains the membership function for average utilization as (see Figure 3.8 (a))

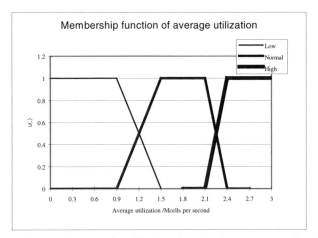

Figure 3.7 (a) The membership function for the average utilization

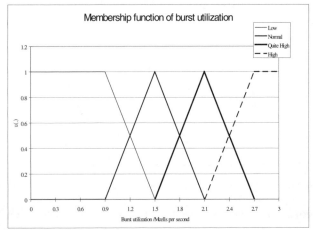

Figure 3.7 (b) The membership function for the burst utilization

$$\mu_{UA,l}(x) = \begin{cases} 1 & 0 \leq x \leq 3 \\ 2.5 - \dfrac{x}{2} & 3 \leq x \leq 5 \\ 0 & 5 \leq x \leq 10 \end{cases}$$

$$\mu_{UA,h}(x) = \begin{cases} 0 & 0 \leq x \leq 7 \\ x - 7 & 7 \leq x \leq 8 \\ 1 & 8 \leq x \leq 10 \end{cases}$$

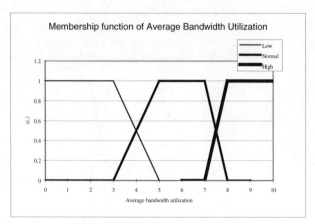

Figure 3.8 (a) The membership function for the average utilization

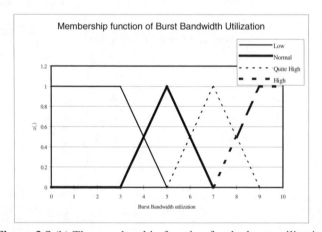

Figure 3.8 (b) The membership function for the burst utilization

$$\mu_{UA,n}(x) = \begin{cases} 0 & 0 \le x \le 3 \\ \dfrac{x}{2} - 1.5 & 3 \le x \le 5 \\ 1 & 5 \le x \le 7 \\ 8 - x & 7 \le x \le 8 \\ 0 & 8 \le x \le 10 \end{cases}$$

and the membership function for burst utilization as (see Figure 3.8 (b))

$$\mu_{UB,l}(x) \;=\; \begin{cases} 1 & 0 \le x \le 3 \\ 2.5 - \dfrac{x}{2} & 3 \le x \le 5 \\ 0 & 5 \le x \le 10 \end{cases}$$

$$\mu_{UB,qh}(x) \;=\; \begin{cases} 0 & 0 \le x \le 5 \\ \dfrac{x}{2} - 2.5 & 5 \le x \le 7 \\ 4.5 - \dfrac{x}{2} & 7 \le x \le 9 \\ 0 & 9 \le x \le 10 \end{cases}$$

$$\mu_{UB,n}(x) \;=\; \begin{cases} 0 & 0 \le x \le 3 \\ \dfrac{x}{2} - 1.5 & 3 \le x \le 5 \\ 3.5 - \dfrac{x}{2} & 5 \le x \le 7 \\ 0 & 7 \le x \le 10 \end{cases}$$

$$\mu_{UB,n}(x) \;=\; \begin{cases} 0 & 0 \le x \le 7 \\ \dfrac{x}{2} - 3.5 & 7 \le x \le 9 \\ 1 & 9 \le x \le 10 \end{cases}$$

Traffic Type

Real-time traffic data (e.g., voice or video) have smaller tolerance to delays. Having the same delay value (to prevent jitter) is more significant to real-time traffic than it does to non-real-time traffic. Therefore, real-time traffic at a node will affect the traffic condition at that node, for which a smaller portion of real-time traffic is desirable. Figure 3.9 shows the chosen membership functions for real-time traffic.

After normalization, one obtains the membership function for real-time traffic as (see Figure 3.10)

$$\mu_R(x) \;=\; \begin{cases} \dfrac{x}{6} & 0 \le x \le 6 \\ 1 & 6 \le x \le 10 \end{cases}$$

The overall performance index is given by

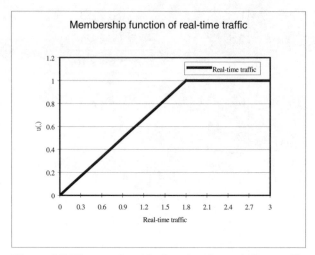

Figure 3.9 The membership function for real-time traffic

$$P_{traffic} =$$

$$\frac{\displaystyle\sum_{j=A,B}\sum_{i=e,n,f}\mu_{Qj,i}(x_{Qj})\cdot x_{Qj} + \sum_{j=A,B}\sum_{i=l,n,h}\mu_{Uj,i}(x_{Uj})\cdot x_{Uj} + \sum\mu_{R}(x_R)\cdot x_R}{\displaystyle\sum_{j=A,B}\sum_{i=e,n,f}\mu_{Qj,i}(x_{Qj}) + \sum_{j=A,B}\sum_{i=l,n,h}\mu_{Uj,i}(x_{Uj}) + \sum\mu_{R}(x_R)}$$

In this evaluation, the smaller the index, the better the performance.

Algorithmic Implementation

The user should enter the parameter matrix as

$$\begin{bmatrix} x_{QA1} & x_{QB1} & x_{UA1} & x_{UB1} & x_{R1} \\ \vdots & \vdots & \vdots & \vdots & \vdots \\ x_{QAn} & x_{QBn} & x_{UAn} & x_{UBn} & x_{Rn} \end{bmatrix}$$

in which each row contains the parameters of the average queue length, burst queue length, average utilization, burst utilization, and real-time traffic at a node of interest.

Then, the user will find the corresponding performance index, given in the form of an output vector:

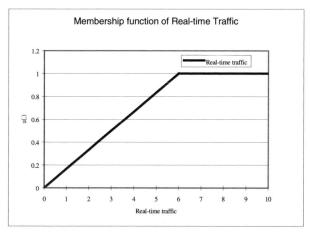

Figure 3.10 The normalized membership function for real-time traffic

$$\begin{bmatrix} p_1 \\ \vdots \\ p_n \end{bmatrix}$$

Performance Evaluation Example

To show one numerical example, suppose that the following raw data set of $n = 12$ nodes are received:

$$a = \begin{bmatrix} 157.6266 & 83.2128 & 1.4568 & 2.3026 & 0.0101 \\ 56.2128 & 59.7598 & 2.8567 & 2.8420 & 2.9460 \\ 44.9573 & 134.4873 & 0.6958 & 2.4399 & 2.6985 \\ 181.7749 & 187.6515 & 1.4360 & 2.7715 & 2.0783 \\ 1.4658 & 68.6295 & 1.5796 & 0.5970 & 1.3190 \\ 117.7479 & 112.5925 & 2.3782 & 2.0228 & 2.1031 \\ 108.4236 & 23.7777 & 0.5790 & 2.7813 & 1.8291 \\ 130.7048 & 33.8043 & 2.7288 & 1.0314 & 0.8997 \\ 62.6870 & 55.7791 & 2.7666 & 1.7835 & 2.5681 \\ 20.0000 & 20.0000 & 0.3000 & 0.3000 & 0.3000 \\ 100.0000 & 100.0000 & 1.5000 & 1.5000 & 1.5000 \\ 200.0000 & 200.0000 & 3.0000 & 3.0000 & 3.0000 \end{bmatrix}$$

This set of data are first normalized and then entered to the algorithm. After computation, one obtains the following result:

$$p = \begin{bmatrix} 6.1348 \\ 6.9229 \\ 5.6839 \\ 7.8848 \\ 2.9542 \\ 6.6394 \\ 4.7817 \\ 4.9465 \\ 5.9301 \\ 1.0000 \\ 5.0000 \\ 10.0000 \end{bmatrix}$$

Conclusion

From this numerical result, one can see that the higher utilization and the longer buffer queue together give a higher performance index. This provides the expected indicator for describing the traffic condition at each node of the network.

Example 3.2 New Car Purchase

The cost of routine maintenance and service charge for a car may gradually become essential for an owner and when it is increasing to an unreasonable level then the owner may want to replace the old car with a new one.

Suppose one is looking for a new car from among four available models, A, B, C, and D. The aim is to choose one model that is cost-effective, in the sense that it is of the lowest price with higher performance and quality features.

Problem Formulation

Suppose that the buyer pre-set three fuzzy membership functions that describe the cost-effectiveness in purchasing a new car, as shown in Figure 3.11 (a)-(c), respectively.

Since these three figures have different dimensions and units, one first needs to normalize them into comparable scales. The three normalized membership functions are shown in Figure 3.12 (a)-(c), which are denoted by $\mu_p(\cdot), \mu_f(\cdot), \mu_c(\cdot)$, respectively.

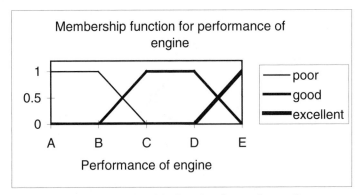

Figure 3.11 (a) Membership function for engine performance

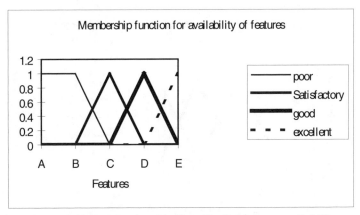

Figure 3.11 (b) Membership function for feature availability

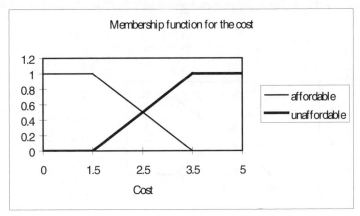

Figure 3.11 (c) Membership function for price

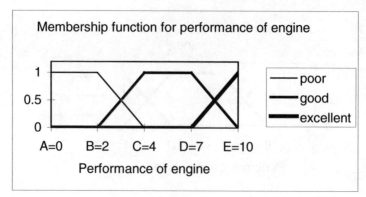

Figure 3.12 (a) Normalized membership function for engine performance

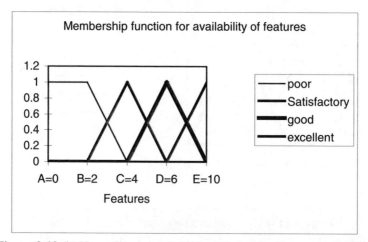

Figure 3.12 (b) Normalized membership function for feature availability

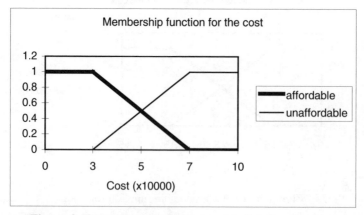

Figure 3.12 (c) Normalized membership function for price

The normalized membership function for performance of engine is given by

$$\mu_{pp}(x_p) = \begin{cases} 1 & 0 \le x_p \le 2 \\ -\dfrac{x_p}{2} + 2 & 2 \le x_p \le 4 \end{cases} \qquad \text{(poor)}$$

$$\mu_{pg}(x_p) = \begin{cases} \dfrac{x_p}{2} - 1 & 2 \le x_p \le 4 \\ 1 & 4 \le x_p \le 7 \\ -\dfrac{x_p}{3} + \dfrac{10}{3} & 7 \le x_p \le 10 \end{cases} \qquad \text{(good)}$$

$$\mu_{pe}(x_p) = \begin{cases} \dfrac{x_p}{3} - \dfrac{7}{3} \end{cases} \qquad 7 \le x_p \le 10 \qquad \text{(excellent)}$$

The normalized membership function for the feature availability is given by

$$\mu_{fp}(x_f) = \begin{cases} 1 & 0 \le x_f \le 2 \\ 2 - \dfrac{x_f}{2} & 2 \le x_f \le 4 \end{cases} \qquad \text{(poor)}$$

$$\mu_{fs}(x_f) = \begin{cases} \dfrac{x_f}{2} - 1 & 2 \le x_f \le 4 \\ 3 - \dfrac{x_f}{2} & 4 \le x_f \le 6 \end{cases} \qquad \text{(satisfactory)}$$

$$\mu_{fg}(x_f) = \begin{cases} \dfrac{x_f}{2} - 2 & 4 \le x_f \le 6 \\ \dfrac{5}{2} - \dfrac{x_f}{4} & 6 \le x_f \le 10 \end{cases} \qquad \text{(good)}$$

$$\mu_{fe}(x_f) = \dfrac{x_f}{4} - \dfrac{3}{2} \qquad 6 \le x_f \le 10 \qquad \text{(excellent)}$$

The normalized membership function for the cost is given by

$$\mu_{fc}(x_c) = \begin{cases} 1 & 0 \le x_c \le 3 \\ -\dfrac{x_c}{4} + \dfrac{7}{4} & 3 \le x_c \le 7 \end{cases} \qquad \text{(affordable)}$$

$$\mu_{fc}(x_c) = \begin{cases} \dfrac{x_c}{4} - \dfrac{3}{4} & 3 \le x_c \le 7 \\ 1 & 7 \le x_c \le 10 \end{cases} \qquad \text{(unaffordable)}$$

Evaluation for the Index of Cost Effectiveness

Because of the lower the cost, the higher the cost-effectiveness, one needs to follow the discussion shown in Fig. 3.4 (c) to re-define the membership function for "price" and accordingly convert x to $10 - x$, namely, $10 \to 0$, $9 \to 1$, ..., $0 \to 10$. For the cases of poor performance with poor/satisfactory features, one should also do the same conversion. The overall index for selecting a car, then, is calculated by the weighted average formula (see the discussions at the end of Section III above).

Implementation

The performance record of the car engine can be obtained from the specification of each model. It is relatively easy to assign an index to this criterion.

The feature availability of a car can also be obtained from the features embedded in the car model, such as car size, interior materials, shape and color of the car body, equipped devices such as stereo radio and CD player, etc. Of course, one may also add some subjective weighing to those models during a drive test.

The cost of each model can be obtained easily from the car dealer who offers that model.

The program is thus readily to be written. As a numerical example, the computed data are shown below:

Index\Model	A	B	C	D
Engine	8.0	7.5	7.0	6.0
Features	7.5	4.8	6.2	6.0
Cost	6.3	5.8	5.5	3.0

The calculated indexes are obtained as below:

Model	A	B	C	D
Index	4.0667	4.1667	4.2333	4.6667

Decision Making

Model D has the higher overall performance index, so it should be chosen as the first candidate when one is purchasing a new car among the four models considered.

Problems

P3.1   Think of an application example, such as performance evaluation, medical diagnosis, product quality control, commercial trading, whatever, that can be performed by using fuzzy logic. It is better to be different from the examples given in this chapter. Describe an approach for designing a computer program, using whatever language, which can perform the task in such a way that when a user input the information data, the program will output the corresponding evaluation result or decision.

P.3.2   Suppose that one is looking for a flat from several available places. The aim is to choose a good one, in the sense that it is of low-price with a large area, short distance to office, good facilities, comfortable environment, nice scenic view and convenient transportation. To reduce the decision complexity, the main factors are:

1)   cost per actual area in terms of dollar per square feet;

2)   location in terms of traveling time between the flat and the office, which is located in a highly active place;

3)   facilities, such as swimming pool and tennis courts;

4)   environment, including view and transportations.

Design an algorithm for decision making, using whatever language, which can perform the task in such a way that when a user input the preference data, the program will output a corresponding suggestion.

CHAPTER 4

Fuzzy Rule Base and Fuzzy Modeling

I. FUZZY RULE BASE

Fuzzy rule base is the core of a fuzzy system. To introduce this concept, a single fuzzy rule is first discussed.

A. Fuzzy IF-THEN Rules

First, recall the fuzzy logic operations *and, or, not, implication*, and also *equivalence*:

$$a \wedge b, \quad a \vee b, \quad \overline{a}, \quad a \Rightarrow b, \quad a \Leftrightarrow b,$$

and their evaluations on a fuzzy set, A, with the membership function $\mu_A(\cdot)$:

$$\mu_A(a \wedge b) := min \{ \mu_A(a), \mu_A(b) \} = \mu_A(a) \wedge \mu_A(b);$$

$$\mu_A(a \vee b) := max \{ \mu_A(a), \mu_A(b) \} = \mu_A(a) \vee \mu_A(b);$$

$$\mu_A(\overline{a}) = \mu_{\overline{A}}(a) = 1 - \mu_A(a);$$

$$\mu_A(a \Rightarrow b) := \mu_A(a) \Rightarrow \mu_A(b) := min \{ 1, 1 + \mu_A(b) - \mu_A(a) \};$$

$$\mu_A(a \Leftrightarrow b) := \mu_A(a) \Leftrightarrow \mu_A(b) := 1 - | \mu_A(a) - \mu_A(b) |.$$

Recall also the fuzzy relations among elements of two fuzzy sets A and B, on which a membership function $\mu_{A \times B}(a,b)$ is defined, with $a \in A$ and $b \in B$. It is clear that one can certainly consider the above fuzzy logic operations as some special fuzzy relations, with $A = B$ and $\mu_{A \times A} = \mu_A$.

In this section, the implication relation $a \Rightarrow b$ and its application in fuzzy logic rules are further studied.

The implication relation $a \Rightarrow b$ can be interpreted, in linguistic terms, as

"IF a is true THEN b is true."

For fuzzy logic performed on a fuzzy set A, there is a membership function μ_A describing the truth values of $a \in A$ and $b \in A$. In this case, a more complete linguistic statement would be

"(IF $a \in A$ is true with a truth value $\mu_A(a)$

THEN $b \in A$ is true with a truth value $\mu_A(b)$)

has a truth value $\mu_A(a \Rightarrow b) = min\ \{\ 1, 1 + \mu_A(b) - \mu_A(a)\ \}$."

In the above, both a and b belong to the same fuzzy subset A and share the same membership function μ_A. If they belong to different fuzzy sets A and B with different membership functions μ_A and μ_B, then one has a nontrivial fuzzy relation, which can be quite complicated. In most cases, however, the implication relation $a \Rightarrow b$, performed on fuzzy sets A and B, where $a \in A$ and $b \in B$, is simply defined in linguistic terms as

"(IF $a \in A$ is true with a truth value $\mu_A(a)$
THEN $b \in B$ is true with a truth value $\mu_B(b)$)
has a truth value $\mu_{A \times B}\ (a \Rightarrow b) = min\ \{\ 1, 1 + \mu_B(b) - \mu_A(a)\ \}$."

Because all such statements have a standard format and their meaning is clear within the context, it is common to write them in the following simple form:

"IF a is A THEN b is B."

A fuzzy logic implication statement of this form is usually called a *fuzzy IF-THEN rule*.

To be more general, let $A_1,...,A_n$, and B be fuzzy subsets with membership functions $\mu_{A_1},...,\mu_{A_n}$, and μ_B, respectively.

Definition 4.1 A *General Fuzzy IF-THEN Rule* has the form

"IF a_1 is A_1 AND ... AND a_n is A_n THEN b is B."

Using the fuzzy logic *AND* operation, this rule is implemented by the following evaluation formula:

$$\mu_{A_1}(a_1) \wedge ... \wedge \mu_{A_n}(a_n) \Rightarrow \mu_B(b),$$

where

$$\mu_{A_i}(a_i) \wedge \mu_{A_j}(a_j) = min\{\ \mu_{A_i}(a_i), \mu_{A_j}(a_j)\ \},\quad 1 \le i, j \le n,$$

and, therefore,

$$\mu_{A_1}(a_1) \wedge \ldots \wedge \mu_{A_n}(a_n) = min\{ \mu_{A_1}(a_1), \ldots, \mu_{A_n}(a_n) \}.$$

About this general fuzzy IF-THEN rule and its evaluation, a few issues have to be clarified:

(i) There is no fuzzy logic *OR* operation in a general fuzzy IF-THEN rule. What should one do if a fuzzy logic implication statement involves the *OR* operation?

(ii) There is no fuzzy logic *NOT* operation in a general fuzzy IF-THEN rule. What should one do if a fuzzy logic implication statement involves the *NOT* operation?

Answers to these questions are given below.

B. Fuzzy Logic Rule Base

Consider questions (i) and (ii). First, consider, for example, the following fuzzy IF-THEN rule containing an *OR* operation:

"IF a_1 is A_1 AND a_2 is A_2 OR a_3 is A_3 AND a_4 is A_4 THEN b is B."

By convention, this is understood in logic as

"(IF a_1 is A_1 AND a_2 is A_2) OR (IF a_3 is A_3 AND a_4 is A_4)

THEN (b is B)."

With this convention and understanding, it is clear that this statement is equivalent to the combination of the following two fuzzy IF-THEN rules:

(1) "IF a_1 is A_1 AND a_2 is A_2 THEN b is B."

(2) "IF a_3 is A_3 AND a_4 is A_4 THEN b is B."

Hence, the fuzzy logic *OR* operation is not necessary to use: it may shorten a statement of a fuzzy IF-THEN rule, but it increases the format complexity of the rules.

Second, consider the fuzzy logic *NOT* operation. For a negative statement like "IF a is not A," one can always interpret it by a positive one "IF $\bar{a}$ is A" or "IF a is $\bar{A}$," where $\bar{A}$ means "not A" in logic theory and also "complement of

A" in set theory. Moreover, the statement "$\overline{a}$ is A" or "a is $\overline{A}$" can be evaluated by

$$\mu_A(\overline{a}) = \mu_{\overline{A}}(a) = 1 - \mu_A(a).$$

Therefore, the fuzzy logic OR operation is not necessary to use either.

Example 4.1

Given a fuzzy logic implication statement

"IF a_1 is A_1 AND a_2 is not A_2 OR a_3 is not A_3 THEN b is B,"

how can one rewrite it as a set of equivalent general fuzzy IF-THEN rules in the unified form?

One may first drop the fuzzy logic OR operation by rewriting the given statement as

    (1) "IF a_1 is A_1 AND a_2 is not A_2 THEN b is B."

    (2) "IF a_3 is not A_3 THEN b is B."

One may then drop the fuzzy logic NOT operation by rewriting them as

    (1') "IF a_1 is A_1 AND $\overline{a}_2$ is A_2 THEN b is B."

    (2') "IF $\overline{a}_3$ is A_3 THEN b is B."

Finally, these two general fuzzy IF-THEN rules can be evaluated as follows:

$$\mu_{A_1}(a_1) \wedge \mu_{A_2}(\overline{a}_2) \Rightarrow \mu_B(b), \quad \text{where } \mu_{A_2}(\overline{a}_2) = 1 - \mu_{A_2}(a_2),$$

$$\mu_{A_3}(\overline{a}_3) \Rightarrow \mu_B(b), \quad \text{where } \mu_{A_3}(\overline{a}_3) = 1 - \mu_{A_3}(a_3).$$

Therefore, one only needs two general fuzzy IF-THEN rules, (1') and (2'), and three membership values $\mu_{A_2}(a_1)$, $\mu_{A_2}(a_2)$, and $\mu_{A_3}(a_3)$, to infer the conclusion "b is B," namely, $b \in B$ with the truth value $\mu_B(b)$ calculated from the three given membership values.

All the other fuzzy logic operations can be simply defined and expressed only by the AND and OR operations. They can be evaluated via the min and max operations as follows:

$$\mu_{A_1}(a_1) \wedge \mu_{A_2}(a_2) = min\{ \mu_{A_1}(a_1), \mu_{A_2}(a_2) \};$$

$$\mu_{A_1}(a_1) \vee \mu_{A_2}(a_2) = max\{ \mu_{A_1}(a_1), \mu_{A_2}(a_2) \};$$

along with

$$\mu_A(\overline{a}) = \mu_{\overline{A}}(a) = 1 - \mu_A(a);$$

$$\mu_A(a \Rightarrow b) = \mu_A(a) \Rightarrow \mu_A(b) = min\{ 1, 1 + \mu_A(b) - \mu_A(a) \};$$

$$\mu_A(a \Leftrightarrow b) = \mu_A(a) \Leftrightarrow \mu_A(b) = 1 - |\mu_A(a) - \mu_A(b)|.$$

Consequently, all finite combinations of these fuzzy logic operations can be expressed only by the *AND* and *OR* operations, so that in any finite fuzzy logic inference statement

IF ... THEN ... ,

where the condition part "IF ..." can be expressed only by the *AND* and *OR* operations.

Consequently, a finite fuzzy logic implication statement can always be described by a set of general fuzzy IF-THEN rules containing only the fuzzy logical *AND* operation, in the following generic form:

(1) "IF a_{11} is A_{11} AND ... AND a_{1n} is A_{1n} THEN b_1 is B_1."

(2) "IF a_{21} is A_{21} AND ... AND a_{2n} is A_{2n} THEN b_2 is B_2."

$\vdots$

(m) "IF a_{m1} is A_{m1} AND ... AND a_{mn} is A_{mn} THEN b_m is B_m."

This family of general fuzzy IF-THEN rules is usually called a *fuzzy logic rule base*.

It is remarked that the number of components in each rule need not be the same. If $n = 2$ but a rule has only one component in the condition part, say,

"IF a_{11} is A_{11} THEN b_1 is B_1,"

one can formally rewrite it as

"IF a_{11} is A_{11} AND a_{12} is I_{12} THEN b_1 is B_1," where I_{12} is a fuzzy subset with $\mu_{I_{12}}(a) = 1$ for all $a \in I_{12}$. Here, one actually inserts an "always true" (redundant) condition into the "IF ... AND ..." part to fill in the gap of the statement. In so doing, the format of a fuzzy logic rule base can be kept simple in a general discussion of the subject.

It is clear that such a general form of a fuzzy logic rule base includes the non-fuzzy case and the unconditional (degenerate) case (with only "b is B") as special cases.

Moreover, this general fuzzy logic rule base (with only the fuzzy logical *AND* operation in the condition part) also covers many unusual fuzzy logic implication statements, such as the one shown in the next example.

Example 4.2

Given a fuzzy logic implication statement

"b is B unless a_1 is A_1 AND ... AND a_n is A_n,"

which is understood in logic as

"(b is B) unless (a_1 is A_1 AND ... AND a_n is A_n),"

one can first convert it by using the fuzzy logic *NOT* and *OR* operations as follows:

"IF $\overline{(a_1 \text{ is } A_1 \text{ AND} ... \text{AND} \, a_n \text{ is } A_n)}$ THEN b is B,"

namely,

"IF a_1 is $\overline{A_1}$ OR ... OR a_n is $\overline{A_n}$ THEN b is B,"

and then replace all the *OR* operations by a fuzzy logic rule base in the following form:

(1) "IF a_1 is $\overline{A_1}$ THEN b is B,"

(2) "IF a_2 is $\overline{A_2}$ THEN b is B,"

$\vdots$

(n) "IF a_n is $\overline{A_n}$ THEN b is B."

This equivalent rule base is in the general format, indeed.

In this example, however, it should be noted that the given statement is not equivalent to the following:

"IF a_1 is A_1 AND ... AND a_n is A_n THEN b is NOT B,"

since the conclusion can be "b has no relation with B."

Finally, note that a fuzzy rule base has to satisfy some properties or requirements (the so-called "3C requirement" – complete, consistent, and concise).

First, a fuzzy rule base has to be **complete,** in the sense that no other possible conditions are left out. The following rule base is incomplete:

(1) IF $a > 0$ THEN $b > 0$,

(2) IF $a = 0$ THEN $b < 0$,

because the case of $a < 0$ is left out.

Second, a fuzzy rule base has to be **consistent,** in the sense that no conclusions are contradictive. The following rule base is inconsistent:

(1) IF $a > 0$ THEN $b > 0$,

(2) IF $a > 0$ THEN $b = 0$,

(3) IF $a = 0$ THEN $b < 0$,

(4) IF $a < 0$ THEN $b = 0$,

since the first two rules contradict each other. Yet this rule base is complete.

Notice that the following two rules are consistent:

(1) IF $a > 0$ THEN $b > 0$,

(2) IF $a = 0$ THEN $b > 0$,

where two different conditions give the same conclusion, which is not a conflict, and the following two rules are consistent, too:

(1) IF $a > 0$ THEN $b > 0$,

(2) IF $a > 0$ THEN $c > 0$,

which are equivalent to "IF $a > 0$ THEN $b > 0$ AND $c > 0$."

Finally, some other requirements may need to be imposed as well for a fuzzy rule base in a particular application. In particular, a rule base should be *concise* with less or no redundancy.

The above 3C requirement provides a guideline for designing a correct, complete and effective fuzzy rule base in engineering applications.

C. Fuzzy IF-THEN Rule Base as a Mathematical Model

There are many different ways to establish a mathematical model for a physical system or process. Normally, a differential or difference equation, or a system of such equations, provides a common approach to such modeling. However, a physical system can also be modeled by means of its input-output relations, or a set of formulas or logical rules, as further explained by the following example, particularly when the system is incorporated with impreciseness, incompleteness, or uncertainties.

Example 4.3

Let

$$y = f(x)$$

be an invertible function defined on $X = [0,4]$ with a value range $Y = [-4,0]$, as shown in Figure 4.1 (a). If a crisp value x is given then $y = f(x)$, and if a crisp value y is given then $x = f^{-1}(y)$.

Suppose that the designer does not actually know the exact formula of f.

Let $\mu_S(\cdot)$, $\mu_M(\cdot)$, and $\mu_L(\cdot)$ be membership functions defined on X and Y, describing "small," "medium," and "large" in absolute values, respectively, as shown in Figure 4.1 (b).

Thus, one may approximate the real function $y = f(x)$ by the following fuzzy rule base, as shown in Figure 4.1 (c):

(1) "IF x is positive small THEN y is negative small."

(2) "IF x is positive medium THEN y is negative medium."

(3) "IF x is positive large THEN y is negative large."

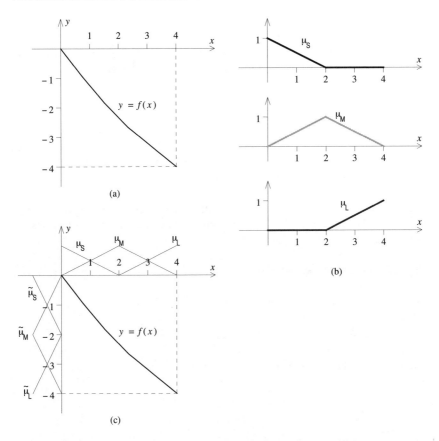

Figure 4.1 An example of approximating a real function by a fuzzy rule base

Using the brief notation "a is A" to mean "$a \in A$ has a membership value $\mu_A(a)$" as before, one may now rewrite the above three implication statements as follows:

(1') "IF x is PS THEN y is NS."

(2') "IF x is PM THEN y is NM."

(3') "IF x is PL THEN y is NL."

Comparing it to the classical two-valued logic inference,

"IF x is positive THEN y is negative"

or

"IF x is small THEN y is small,"

it can be seen that the following long statement from the classical logic is needed to express the same meaning of the fuzzy logic inference (1):

"IF x is positive AND x is small THEN y is negative AND y is small."

Even if so, the classical logic can only be used to determine "x is small" or "x is not small," while the fuzzy membership function μ_S gives infinitely many different truth values to describe how small x is.

More about fuzzy modeling will be further discussed in the next section, Section II. To prepare for it, the important issue of evaluating fuzzy rules is first addressed.

D. Evaluation of Fuzzy IF-THEN Rules

Consider the problem of evaluating a fuzzy IF-THEN rule:

$$\mu_{A \Rightarrow B}(a,b) \quad \text{namely} \quad \mu_A(a) \Rightarrow \mu_B(b), \quad a \in A, \, b \in B.$$

For the classical two-valued logic, this evaluation is simple:

$$\mu_B(b) = \begin{cases} 1 & \text{if } \mu_A(a)=1, \\ 0 & \text{if } \mu_A(a)=0, \end{cases}$$

namely, "$a \in A \Rightarrow b \in B$," and

$$\mu_{\bar{B}}(b) = \begin{cases} 1 & \text{if } \mu_{\bar{A}}(a)=1, \\ 0 & \text{if } \mu_{\bar{A}}(a)=0, \end{cases}$$

that is, "$a \notin A \Rightarrow b \notin B$."

For fuzzy logic, however, according to the choice of a particular logical system, there are several options for the IF-THEN rule

"$\mu_A(a) \Rightarrow \mu_B(b)$"

such as

(a) $\mu_{A \Rightarrow B}(a,b) = min\{ \mu_A(a), \mu_B(b) \}$;

(b) $\mu_{A \Rightarrow B}(a,b) = \mu_A(a) \cdot \mu_B(b)$;

(c) $\mu_{A \Rightarrow B}(a,b) = min\{\ 1,\ 1 + \mu_B(b) - \mu_A(a)\ \}$;

(d) $\mu_{A \Rightarrow B}(a,b) = max\{\ min\{\ \mu_A(a),\ \mu_B(b)\ \},\ 1 - \mu_A(a)\ \}$;

(e) $\mu_{A \Rightarrow B}(a,b) = max\{\ 1 - \mu_A(a),\ \mu_B(b)\ \}$;

(f) Goguen's formula:

$$\mu_{A \Rightarrow B}(a,b) = \begin{cases} 1 & \text{if } \mu_A(a) \le \mu_B(b), \\ \dfrac{\mu_B(b)}{\mu_A(a)} & \text{if } \mu_A(a) > \mu_B(b). \end{cases}$$

All these evaluation formulas are valid for the fuzzy logic inference purpose, provided that one uses consistently the same formula for the implication relation $\Rightarrow$. Of course, different formulas give different resulting values, which merely imply different degrees of inference based on different logical systems, but not the validity of the answers. Obviously, formulas (a) and (b) are very simple to use; but they are the same as the logical *AND* operation $\wedge$, formulas (d) and (f) seem too complicated. The most common one in applications is formula (c), which has been used and will be continuously used throughout this text.

A final remark is that for the following general fuzzy IF-THEN rule:

"IF a_1 is A_1 AND ... AND a_n is A_n THEN b is B,"

one can first evaluate the condition part by

$$\mu_A(a_1,...,a_n) = min\{\ \mu_{A_1}(a_1),...,\ \mu_{A_n}(a_n)\ \}\ ,$$

and then evaluate $\mu_A(a_1,...,a_n) \Rightarrow \mu_B(b)$ by

$$\mu_{A \Rightarrow B}(a_1,...,a_n,b) = min\{\ 1,\ 1 + \mu_B(b) - \mu_A(a_1,...,a_n)\ \}\ .$$

II. FUZZY MODELING

Modeling, in a general sense, refers to the establishment of a description of a physical system (a plant, a process, etc.) in mathematical terms, which characterizes the input-output behavior of the underlying system.

To describe a physical system, such as a circuit or a microprocessor, one has to use a mathematical formula or equation that can represent the system both qualitatively and quantitatively. Such a formulation is, by nature, a mathematical representation, called a *mathematical model*, of the physical system in interest.

Most physical systems, particularly those complex ones, are extremely difficult to model by an accurate and precise mathematical formula or equation due to the complexity of the system structure, nonlinearity, uncertainty, randomness, etc. Therefore, *approximate modeling* is often necessary and practical in real-world applications.

Intuitively, approximate modeling is always possible. However, the key questions are what kind of approximation is good, where the sense of "goodness" has to be first defined, of course, and how to formulate such a good approximation in modeling a system such that it is mathematically rigorous and can produce satisfactory results in both theory and applications.

From the detailed studies in the last three chapters, it is clear that interval mathematics and fuzzy logic together can provide a promising alternative to mathematical modeling for many physical systems that are too vague or too complicated to be described by simple and crisp mathematical formulas or equations. When interval mathematics and fuzzy logic are employed, the interval of confidence and the fuzzy membership functions are used as approximation measures, leading to the so-called *fuzzy systems modeling*.

Following the traditional classification in the field of control systems, a system that describes the input-output behavior in a way similar to a mathematical mapping without involving a differential operator or equation is called a *static system*. In contrast, a system described by a differential operator or equation is called a *dynamic system*. In this section, both static and dynamic fuzzy systems modeling are discussed.

A. Basic Concept of Systems Modeling

Suppose that one has an unknown system ("black box"), for which only a set of its inputs $x_1, ..., x_n$ and outputs $y_1, ..., y_m$ can be measured (or observed) and these data are available to use.

A mathematical description is needed to qualitatively and quantitatively characterize this unknown system, in the sense that by inputting $x_1, ..., x_n$ into the mathematical description one can always obtain the corresponding outputs $y_1, ..., y_m$. Establishing such a mathematical description is called *mathematical modeling* for the unknown system (Figure 4.2).

Figure 4.2 An unknown system as a "black box"

As usual, the mathematical description can be a mathematical formula, such as a mapping or a functional that relates the inputs to the outputs in the form

$$\begin{cases} y_1 = f_1(x_1,...,x_n), \\ \vdots \\ y_m = f_m(x_1,...,x_n); \end{cases} \tag{4.1}$$

or a set of differential equations in the form

$$\begin{cases} y_1 = g_1(x_1,...,x_n,\dot{x}_1,...,\dot{x}_n), \\ \vdots \\ y_m = g_m(x_1,...,x_n,\dot{x}_1,...,\dot{x}_n); \end{cases} \tag{4.2}$$

or a logical linguistic statement in the form

IF (input x_1) AND ... AND (input x_n)
THEN (output y_1) AND ... AND (output y_m). (4.3)

Fuzzy systems modeling is to quantify the logical linguistic form (4.3) by using fuzzy logic and the mathematical functional model (4.1) or by using fuzzy logic with the differential equation model (4.2). The result of the former is a *static fuzzy system,* and of the latter, a *dynamic fuzzy system.*

B. Modeling of Static Fuzzy Systems

Return to the "black box" shown in Figure 4.2. Since all the inputs x_1, ..., x_n and outputs y_1, ..., y_m are assumed to be available, the logical linguistic statement (4.3) has actually described the unknown system, on the basis of the available data. Yet this is not the ultimate purpose of mathematical modeling, since if a new input x_{n+1} comes in, one does not know what the corresponding output should be. The main purpose of mathematical modeling, therefore, is not only to correctly describe the existing input-output relations through the unknown system but also to enable the established model to approximately describe other possible hidden input-output relations of the system. Thus, a general approach is to first quantify the linguistic statement (4.3) and then to relay the quantified logical input-output relations by using the mathematical functions (4.1), or differential equations (4.2).

Recall that a finite fuzzy logic implication statement can always be described by a set of general fuzzy IF-THEN rules containing only the fuzzy logic AND operation, in the following multi-input single-output form:

(1) "IF (x_1 is X_{11}) AND ... AND (x_n is X_{1n}) THEN (y is Y_1)."

(2) "IF (x_1 is X_{21}) AND ... AND (x_n is X_{2n}) THEN (y is Y_2)."

$\vdots$

(N) "IF (x_1 is X_{N1}) AND ... AND (x_n is X_{Nn}) THEN (y is Y_N)."

Here, the phrase "x is X" is an abbreviation of the complete statement "x belongs to the fuzzy set X with a corresponding membership value $\mu_X(x)$." In the following, X_{11}, ..., X_{Nn} and Y_1, ..., Y_N are all closed intervals.

For simplicity of discussion, first consider the case where $N = 1$ with only one fuzzy IF-THEN rule:

"IF (x_1 is X_1) AND ... AND (x_n is X_n) THEN (y is Y)."

An example is the following rule with constants $\{a_0,a_1,...,a_n\}$:

R^1:   IF (x_1 is X_1) AND ... AND (x_n is X_n)
           THEN $y = a_0 + a_1x_1 + ... + a_nx_n$.

When a set of particular inputs are available:

$$x_1 = x_1^0 \in X_1, ..., x_n = x_n^0 \in X_n, \text{ with } \mu_{X_1}(x_1^0), ..., \mu_{X_n}(x_n^0),$$

the output y will assume the value

$$y^0 = a_0 + a_1 x_1^0 + ... + a_n x_n^0, \tag{4.4}$$

with membership value given by the *general rule* (see Chapter 1, Section IV.B), as

$$\mu_Y(y^0) = \bigvee_{y^0 = a_0 + ... + a_n x_n^0} \{\mu_{X_1}(x_1^0) \wedge ... \wedge \mu_{X_n}(x_n^0)\}. \tag{4.5}$$

If there are more than one fuzzy IF-THEN rule (N>1):

$$R^i: \quad \text{IF } (x_1 \text{ is } X_{i1}) \text{ AND } ... \text{ AND } (x_n \text{ is } X_{in})$$
$$\text{THEN } y_i = a_{i0} + a_{i1}x_1 + ... + a_{in}x_n, \quad i=1,...,N,$$

then, with the given set of inputs $x_1 = x_1^0 \in X_1, ..., x_n = x_n^0 \in X_n$, namely, the same inputs applied to all the different rules, one will have

$$y_1^0 = a_{10} + a_{11} x_1^0 + ... + a_{1n} x_n^0,$$
$$y_2^0 = a_{20} + a_{21} x_1^0 + ... + a_{2n} x_n^0,$$
$$\vdots \qquad\qquad\qquad\qquad\qquad (4.6)$$
$$y_N^0 = a_{N0} + a_{N1} x_1^0 + ... + a_{Nn} x_n^0.$$

The corresponding membership values for these outputs are given as

$$\mu_Y(y_i^0) = \bigvee_{y_i^0 = a_{i0}+...+a_{in}x_n^0} \{\mu_{X_1}(x_1^0)\wedge...\wedge\mu_{X_n}(x_n^0)\}, \quad i=1, ..., N. \quad (4.7)$$

In a typical modeling approach, the final single output, y, is usually obtained via the following *weighted average formula*:

$$y = \frac{\sum_{i=1}^{N}\mu_Y(y_i^0)\cdot y_i^0}{\sum_{i=1}^{N}\mu_Y(y_i^0)}, \qquad (4.8)$$

where "·" is the ordinary algebraic multiplication.

For the most general situation, assume that all the coefficients $\{a_{i0},a_{i1},...,a_{in} \mid i=1,...,N \}$ are uncertain and belong to certain intervals:

$$a_{i0} \in A_0, ..., a_{in} \in A_n, \qquad i=1,...,N,$$

where, for example,

$$A_0 = [\, min\{a_{10},a_{20},...,a_{N0}\}, max\{ a_{10},a_{20},...,a_{N0}\} \,],$$
$$A_1 = [\, min\{ a_{11},a_{21},...,a_{N1}\}, max\{ a_{11},a_{21},...,a_{N1}\} \,],$$
$$\vdots$$

$$A_n = [\ min\{a_{1n},a_{2\,n},...,a_{Nn}\ \},\ max\{\ a_{1n},a_{2\,n},...,a_{Nn}\ \}\ ].$$

Thus, with the given inputs $x_1 \in X_1$, $x_2 \in X_2$, ..., $x_n \in X_n$, the output becomes

$$Y = A_0 + A_1 \cdot X_1 + ... + A_n \cdot X_n, \tag{4.9}$$

which yields the fuzzy subset (interval) for y, with the membership functions given by the general rule as

$$\mu_{Y,i}(y_i) = \bigvee_{y_i = a_{i0}+...+a_{in}x_n} \{\mu_{X_1}(x_1^0) \wedge ... \wedge \mu_{X_n}(x_n^0)\},\ i = 1, ..., N. \tag{4.10}$$

Finally, the output y is computed by the weighted average formula, as

$$y = \frac{\sum\limits_{i=1}^{N} \mu_{Y,i}(y_i) \cdot y_i}{\sum\limits_{i=1}^{N} \mu_{Y,i}(y_i)} = \sum_{i=1}^{N} \beta_i \cdot y_i \tag{4.11}$$

where

$$\beta_i = \frac{\mu_{Y,i}(y_i)}{\sum\limits_{i=1}^{N} \mu_{Y,i}(y_i)}.$$

This is a convex combination of the outputs y_i, $i=1,...,N$.

The three formulas (4.6)-(4.8) or (4.9)-(4.11) are sometimes called the *input-output algorithm* for static fuzzy system modeling under the fuzzy IF-THEN rules R^i, $i = 1, ..., N$, where the former has constant coefficients while the latter has interval coefficients.

Example 4.4

Consider an unknown system with two inputs, x_1, x_2, and one output, y, as shown in Figure 4.3 (a).

Let inputs x_1 be in the range $X_1 = [0,20]$ and x_2 in $X_2 = [0,10]$. Suppose that X_1 has two associate membership functions, $\mu_{S_1}(\cdot)$ and $\mu_{L_1}(\cdot)$, describing "small" and "large," respectively, and similarly X_2 has $\mu_{S_2}(\cdot)$ and $\mu_{L_2}(\cdot)$, as shown in Figure 4.3 (b)-(c).

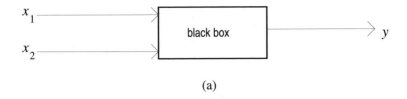

(a)

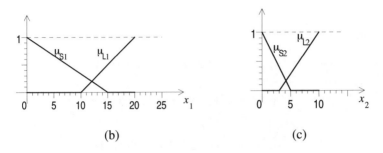

(b)   (c)

Figure 4.3 The unknown system and its two inputs in Example 4.4

Now, suppose that from experiments the following input-output relations (fuzzy IF-THEN implication rules) are obtained:

R^1:   IF x_1 is small AND x_2 is small THEN $y = x_1 + x_2$

R^2:   IF x_1 is large THEN $y = \frac{1}{2} x_1$

R^3:   IF x_2 is large THEN $y = \frac{1}{3} x_2$

Otherwise, $y = 0$.

Here, assume that x_1 and x_2 will not be both large in this example, just for simplicity of discussion with consistence.

Let $x_1^0 = 13$ and $x_2^0 = 4$ be given, for instance. It can be found from Figures 4.3 (b)-(c) that

$$\mu_{S_1}(x_1^0) = 2/15, \ \mu_{L_1}(x_1^0) = 3/10, \ \mu_{S_2}(x_2^0) = 1/5, \ \mu_{L_2}(x_2^0) = 1/7.$$

Then, following the input-output algorithm, compute the following:

(i) The corresponding outputs:

$$y_1^0 = x_1^0 + x_2^0 = 13 + 4 = 17,$$

$$y_2^0 = \frac{1}{2} x_1^0 = \frac{13}{2},$$

$$y_3^0 = \frac{1}{3} x_2^0 = \frac{4}{3}.$$

(ii) The fuzzy set Y:

$$a_{11} = 1, a_{12} = 1, a_{21} = 1/2, a_{22} = 0, a_{31} = 0, a_{32} = 1/3$$

$$\Rightarrow \qquad A_1 = [\, min\{1, \tfrac{1}{2}, 0\}, max\{1, \tfrac{1}{2}, 0\}\,] = [0,1],$$

$$A_2 = [\, min\{1, 0, \tfrac{1}{3}\}, max\{1, 0, \tfrac{1}{3}\}\,] = [0,1],$$

$$Y = A_1 \cdot X_1 + A_2 \cdot X_2 = [0,1] \cdot [0,20] + [0,1] \cdot [0,10] = [0,30].$$

(iii) The fuzzy membership values of the outputs:

$$\mu_{Y,1}(y_1^0) = \bigvee_{y_1^0 = x_1^0 + x_2^0} \{\, \mu_{S_1}(x_1^0) \wedge \mu_{S_2}(x_2^0) \,\}$$

$$= max\{\, min\{\, \mu_{S_1}(x_1^0), \mu_{S_2}(x_2^0) \,\} \,\}$$

$$= max\{\, min\,\{\, \tfrac{2}{15}, \tfrac{1}{5} \,\} \,\}$$

$$= max\{\, \tfrac{2}{15} \,\}$$

$$= \frac{2}{15},$$

$$\mu_{Y,2}(y_2^0) = \bigvee_{y_1^0 = \frac{1}{2} x_1^0} \{\, min\,\{\, \mu_{L_1}(x_1^0) \,\} \,\} = max\{\, \tfrac{3}{10} \,\}$$

$$= \frac{3}{10},$$

$$\mu_{Y,3}(y_3^0) = \bigvee_{y_1^0 = \frac{1}{3}x_2^0} \{ min \{ \mu_{L_2}(x_2^0) \} \} = max\{\frac{1}{7}\} = \frac{1}{7}.$$

(iv) The average value of the final output, corresponding to the inputs $x_1^0 = 13$ and $x_2^0 = 4$:

$$y^0 = \frac{\sum_{i=1}^{3}\mu_{Y,i}(y_i^0)\cdot y_i^0}{\sum_{i=1}^{3}\mu_{Y,i}(y_i^0)} = \frac{\frac{2}{15}\cdot 17 + \frac{3}{10}\cdot\frac{13}{2} + \frac{1}{7}\cdot\frac{4}{3}}{\frac{2}{15} + \frac{3}{10} + \frac{1}{7}} \approx 7.649.$$

(v) The three membership functions, one "small" and two "large," for the final output:

$$\mu_S(y) = \begin{cases} -\frac{y}{20}+1, & 0 \le y \le 20, \\ 0, & 20 \le y \le 30, \end{cases}$$

$$\mu_{L_1}(y) = \begin{cases} 0, & 0 \le y \le 5, \\ \frac{y}{5}-1, & 5 \le y \le 10, \\ 1 & 10 \le y \le 30, \end{cases}$$

$$\mu_{L_2}(y) = \begin{cases} 0, & 0 \le y \le 1, \\ \frac{3y}{7}-\frac{3}{7}, & 1 \le y \le \frac{10}{3}, \\ 1 & \frac{10}{3} \le y \le 30, \end{cases}$$

in which $\mu_S(y)$ can be computed from μ_{S_1} and μ_{S_2} by the α-cut operations, and, similarly, $\mu_{L_1}(y)$ from $\mu_{L_1}(x_1^0)$ and $\mu_{L_2}(y)$ from $\mu_{L_2}(x_2^0)$, respectively.

Here, to find $\mu_{L_1}(y)$ from $\mu_{L_1}(x_1^0)$, for instance, one uses α-cut on the membership function curve $\mu_{L_1}(x_1)$: $\alpha = (x - 10)/10$, and obtain $x_{11} = 10\alpha + 10$, while $x_{12} = 20$ is obtained from Figure 4.3 (b). Thus, one has $[x_{11},x_{12}] = [10\alpha+10,20]$. It then follows from the implication rule R^2 ($y_2=\frac{1}{2}x_1$) that the output interval is $[y_{21},y_{22}] = [5\alpha+5,10]$. Then, by setting $y_{21} = 5\alpha + 5$, one has $\alpha = y_{21}/5 - 1$ on $[5,10]$ (since $\mu_{L_1}(y =5) = 0$ and $\mu_{L_1}(y =10) = 1$). These membership functions are finally extended to the interval $Y = [0,30]$, which

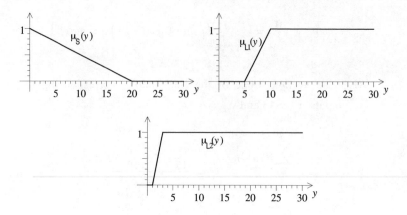

Figure 4.4 The three output membership functions in Example 4.4

was determined above. The resulting three output membership functions are shown in Figure 4.4.

C. Parameters Identification in Static Fuzzy Modeling

Note that the approach discussed above, including the input-output algorithm and Example 4.4 therein, is about how to describe the input-output relations for an unknown system, where the input-output relations are partially known; namely, the constant coefficients in the relations

$$y_i = a_{i0} + a_{i1}x_1 + \ldots + a_{in}x_n, \ x_i \in X_i, \ i=1,\ldots,N,$$

are all given. In general, one has to identify the unknown coefficients (system parameters) based on the available input-output data under a certain meaningful model-matching criterion such as least-squares. This is *system parameter identification*.

In order to solve the system modeling problem, one has to determine the following five items by using the available input-output data:

(i) $x_1,\ldots,x_n$: input variables, used as the premises of fuzzy logic implications.

(ii) $X_1,\ldots,X_n$: input variable intervals, used as fuzzy sets.

(iii) $\mu_{X_1},\ldots,\mu_{X_n}$: membership functions of the input variables, used to measure the qualities and quantities of the inputs.

(iv) R^i: relations (implications), used as descriptions of system input-output behavior, which take the form

$$y_i = a_{i0} + a_{i1}x_1 + ... + a_{in}x_n, \; x_i \in X_i, \; i = 1,...,N.$$

(v) $a_{i0},...,a_{in}$ ($i=1,...,N$): constant parameters of the model, used for the overall mathematical model.

Once these steps have been carried out, the fuzzy system modeling can be completed by applying the input-output algorithm given above.

Next, the least-squares method is applied to determine optimal constant parameters for the static fuzzy modeling problem discussed above.

The Least-Squares Parameter Identification Scheme:

Start with the fuzzy IF-THEN rules that provide the input-output relations (implications):

R^1:   IF x_1 is X_{11} AND ... AND x_n is X_{1n}
           THEN $y_1 = a_{10} + a_{11}x_1 + ... + a_{1n}x_n$,

$\vdots$

R^N: IF x_1 is X_{N1} AND ... AND x_n is X_{Nn}
           THEN $y_N = a_{N0} + a_{N1}x_1 + ... + a_{Nn}x_n$,

where, for each $i = 1,...,N$, all X_{ij} share the same fuzzy subset X_j and the same membership function $\mu_{X_j}, j = 1,...,n$, and

$$a_{i0} \in A_0, \; a_{i1} \in A_1, \; ..., \; a_{in} \in A_n.$$

The final output y for the inputs $x_1,...,x_n$ is given by formulas (4.10)-(4.11):

$$y = \frac{\sum_{i=1}^{N} \left(\mu_{X_{i1}}(x_1) \wedge ... \wedge \mu_{X_{in}}(x_n) \right) \cdot \left(a_{i0} + a_{i1}x_1 + ... + a_{in}x_n \right)}{\sum_{i=1}^{N} \left(\mu_{X_{i1}}(x_1) \wedge ... \wedge \mu_{X_{in}}(x_n) \right)}, \tag{4.12}$$

where $\mu_{X_{ij}}(x_j) = \mu_{X_j}(x_j)$ is the membership value of x_j obtained in the rule R^i, $i = 1,...,N$, $j = 1,...,n$. Let

$$\beta_i = \frac{\mu_{X_{i1}}(x_1) \wedge \ldots \wedge \mu_{X_{in}}(x_n)}{\displaystyle\sum_{i=1}^{N} \left(\mu_{X_{i1}}(x_1) \wedge \ldots \wedge \mu_{X_{in}}(x_n) \right)}, \quad i=1,\ldots,N, \qquad (4.13)$$

and rewrite (4.12) as

$$y = \sum_{i=1}^{N} \beta_i (a_{i0} + a_{i1}x_1 + \ldots + a_{in}x_n)$$

$$= \sum_{i=1}^{N} \begin{bmatrix} \beta_i & \beta_i x_1 & \cdots & \beta_i x_n \end{bmatrix} \begin{bmatrix} a_{i0} \\ a_{i1} \\ \vdots \\ a_{in} \end{bmatrix}. \qquad (4.14)$$

Suppose that a set of input-output data is given:

$$x_1^1, \quad x_2^1, \quad \ldots, \quad x_n^1, \quad y^1;$$

$$x_1^2, \quad x_2^2, \quad \ldots, \quad x_n^2, \quad y^2;$$

$$\vdots$$

$$x_1^M, \quad x_2^M, \quad \ldots, \quad x_n^M, \quad y^M.$$

To determine the coefficients $\{a_{i0}, a_{i1}, \ldots, a_{in} | \ i = 1,\ldots,N \ \}$ by using this set of data, the standard least-squares method is applied. By substituting the data set into (4.14) successively, one obtains a system of algebraic equations:

$$\Lambda\theta = b, \qquad (4.15)$$

where

$$\Lambda = \Lambda_{M\times(N\times(n+1))}$$

$$= \begin{bmatrix} \beta_{11} & \cdots & \beta_{N1} & x_1^1\beta_{11} & \cdots & x_1^1\beta_{N1} & \cdots & x_n^1\beta_{11} & \cdots & x_n^1\beta_{N1} \\ \vdots & & \vdots & \vdots & & \vdots & & \vdots & & \vdots \\ \beta_{1M} & \cdots & \beta_{NM} & x_1^M\beta_{1M} & \cdots & x_1^M\beta_{NM} & \cdots & x_n^M\beta_{1M} & \cdots & x_n^M\beta_{NM} \end{bmatrix},$$

$$\theta = \begin{bmatrix} a_{10} & \cdots & a_{N0} & a_{11} & \cdots & a_{N1} & \cdots & a_{1n} & \cdots & a_{Nn} \end{bmatrix}^{\mathrm{T}}_{(N \times (n+1)) \times 1},$$

$$b = \begin{bmatrix} y^1 & y^2 & \cdots & y^M \end{bmatrix}^{\mathrm{T}}_{M \times 1},$$

$$\beta_{kj} = \frac{\mu_{X_{k1}}(x_1^j) \wedge \ldots \wedge \mu_{X_{kn}}(x_n^j)}{\sum\limits_{k=1}^{N} \left(\mu_{X_{k1}}(x_1^j) \wedge \ldots \wedge \mu_{X_{kn}}(x_n^j) \right)}, \quad k = 1,\ldots,N; \ j = 1,\ldots,M.$$

It then follows from the standard least-squares method that the optimal coefficients θ^* are given by

(i)   if Λ has a full column-rank:

$$\theta^* = [\Lambda^{\mathrm{T}} \Lambda]^{-1} \Lambda^{\mathrm{T}} b ; \tag{4.16}$$

(ii) if Λ does not have a full column-rank, then decompose $\Lambda = LC$, where L has a full column-rank and C has a full row-rank; both need not be square, which is always possible, and

$$\theta^* = C^{\mathrm{T}} [CC^{\mathrm{T}}]^{-1} [L^{\mathrm{T}} L]^{-1} L^{\mathrm{T}} b. \tag{4.17}$$

This is the standard pseudo-inverse solution of the least-squares optimization.

Example 4.5

Consider an unknown system with one input x and one output y, as shown in Figure 4.5 (a). Let the input x be within the range $X = [0,10]$, with the associate membership functions $\mu_S(\cdot)$ and $\mu_L(\cdot)$ describing "small" and "large," respectively, as shown in Figure 4.5 (b)-(c).

Suppose that the unknown system is the following precise fuzzy static system:

The real system:

R_0^1 :   IF x is small with a membership value $\mu_S(x)$
            THEN $y = 2 + 0.5x$.

R_0^2 :   IF x is large with a membership value $\mu_L(x)$
            THEN $y = 9 + 0.3x$.

Assume that this real system is only used for generating data, but is unknown to the user in this example of system parameter identification.

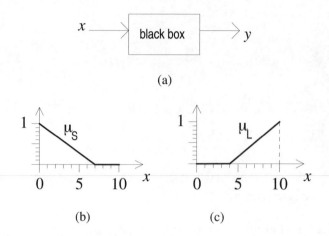

(a)

(b)   (c)

Figure 4.5 The unknown system and its two input membership functions

Suppose, on the other hand, that one tries to identify this system by the following estimated *fuzzy static model*:

R^1:   IF x is small with a membership value $\tilde{\mu}_S(x)$
              THEN $y = a_{10} + a_{11}x$.

R^2:   IF x is large with a membership value $\tilde{\mu}_L(x)$
              THEN $y = a_{20} + a_{21}x$.

Here, in general, $\tilde{\mu}_S$ and $\tilde{\mu}_L$ are likely different from μ_S and μ_L, respectively, since one does not exactly know the real system. For simplicity of discussion in this example, however, suppose that one can set the membership functions in the modeling phase in such a way that $\tilde{\mu}_S = \mu_S$ and $\tilde{\mu}_L = \mu_L$, but does not know those coefficients a_{10}, a_{11}, a_{20}, and a_{21}.

Now, suppose that one has carried out some experiments on the unknown system and obtained the following noisy data (see Figure 4.6):

Suppose also that in this example, one created the data by using the true input-output relation R_0^1 and R_0^2 in the real system, namely, $y = 2 + 0.5x$ and $y = 9 + 0.3x$, respectively, by incorporating with pseudo Gaussian white noise of zero mean and a small covariance $\sigma = 0.01$.

To follow the least-squares parameter identification scheme, one first computes the following (see Figure 4.5 (b)-(c) and Table 4.1):

Table 4.1 Input-Output Data

X	0.2	0.4	0.6	0.8	1.0	...	10.0
Y	1.6	2.8	3.5	1.5	3.0	...	10.4

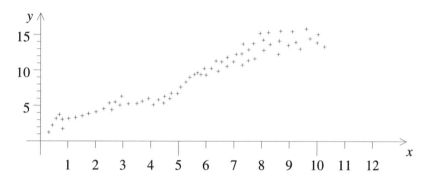

Figure 4.6 Experimental input-output data for Example 3.5

$$\mu_S(x_1) \quad = \mu_S(0.2) = 0.97,$$

$$\mu_S(x_2) \quad = \mu_S(0.4) = 0.94,$$

$$\vdots$$

$$\mu_S(x_{50}) = \mu_S(10.0) = 0.00,$$

and

$$\mu_L(x_1) = \mu_L(0.2) = 0.00,$$

$$\vdots$$

$$\mu_L(x_{49}) = \mu_L(9.8) = 0.97,$$

$$\mu_L(x_{50}) = \mu_L(10.0) = 1.00.$$

Then, compute:

$$\beta_{1,1} = \frac{\mu_S(x_1)}{\mu_S(x_1)+\mu_L(x_1)}, \qquad \beta_{2,1} = \frac{\mu_L(x_1)}{\mu_S(x_1)+\mu_L(x_1)},$$

$$\beta_{1,2} = \frac{\mu_S(x_2)}{\mu_S(x_2)+\mu_L(x_2)}, \qquad \beta_{2,2} = \frac{\mu_L(x_2)}{\mu_S(x_2)+\mu_L(x_2)},$$

$$\vdots \qquad\qquad\qquad \vdots$$

$$\beta_{1,50} = \frac{\mu_S(x_{50})}{\mu_S(x_{50})+\mu_L(x_{50})}, \qquad \beta_{2,50} = \frac{\mu_L(x_{50})}{\mu_S(x_{50})+\mu_L(x_{50})}.$$

Finally, solve the following equation by using the least-squares formula (4.16) or (4.17):

$$\Lambda\theta = b,$$

where, in this example,

$$\Lambda = \begin{bmatrix} \beta_{1,1} & \beta_{2,1} & x_1\beta_{1,1} & x_1\beta_{2,1} \\ \vdots & \vdots & \vdots & \vdots \\ \beta_{1,50} & \beta_{2,50} & x_{50}\beta_{1,50} & x_{50}\beta_{2,50} \end{bmatrix}_{50\times4},$$

$$\theta = \begin{bmatrix} a_{10} & a_{20} & a_{11} & a_{21} \end{bmatrix}^{\mathrm{T}}_{4\times1},$$

$$b = \begin{bmatrix} y_1 & y_2 & \cdots & y_{50} \end{bmatrix}^{\mathrm{T}}_{50\times1},$$

in which x_k and y_k, $k = 1, 2, ..., 50$, are given in Table 4.1 (or Figure 4.6). The result is obtained from the given simulation data, as

$$\theta^* = [\,\Lambda^{\mathrm{T}}\Lambda\,]^{-1}\,\Lambda^{\mathrm{T}}b = \begin{bmatrix} 1.87 \\ 8.68 \\ 0.56 \\ 0.38 \end{bmatrix}.$$

Thus, the identified model, which is optimal in the sense of least-squares, is obtained as follows:

R^1:   IF x is small with a membership value $\tilde{\mu}_S(x)$
                THEN $y = 1.87 + 0.56x$.

R^2: IF x is large with a membership value $\tilde{\mu}_L(x)$
THEN $y = 8.68 + 0.38x$.

The associate membership functions are $\tilde{\mu}_S = \mu_S$ and $\tilde{\mu}_L = \mu_L$, as shown in Figure 4.5 (b)-(c).

D.* Discrete-Time Dynamic Fuzzy Systems Stability

A static fuzzy model in the following form was established above:

R^i: IF x_1 is X_{i1} AND ... AND x_n is X_{in}
THEN $y_i = a_{i0} + a_{i1}x_1 + ... + a_{in}x_n$, $i=1,...,N$,   (4.18)

and the final output is

$$ y = \frac{\sum_{i=1}^{N} \mu_{Y,i}(y_i) \cdot y_i}{\sum_{i=1}^{N} \mu_{Y,i}(y_i)} , \qquad i=1,...,N. \qquad (4.19) $$

If the input-output relations are such that

$x_1 = x(k)$,

$x_2 = x(k-1)$,

$$ \vdots $$

$x_n = x(k-(n-1))$,

and

$y_i = y_i(k)$, $i=1,...,N$,

so that

$y_i(k) = a_{i0} + a_{i1}x(k) + a_{i2}x(k-1) + ... + a_{in}x(k-(n-1))$, $i=1,...,N$,
(4.20)

then this new system is considered to be *dynamic*, since it is described by a set of difference equations rather than a simple input-output mapping.

This case is special in that all the inputs $x_1,..., x_n$ are related (each state is the delay of its previous one) and the output is given at the last step of the process.

Equation (4.20) is a typical discrete-time dynamic system in the classical systems theory if all quantities involved are crisp.

D.1. Dynamic Fuzzy Systems without Control

It is particularly important to study the stability of a discrete-time dynamic (fuzzy) system described by (4.20). The reason is that the resulting model should be stable so as to work well in the sense that it not only approximates the unknown system closely (preferably, optimal in some sense) but also performs its actions stably, particularly for dynamic processes.

Definition 4.1 A typical *single-input/single-output* (SISO), *discrete-time*, *dynamic fuzzy system* is a fuzzy model described by a set of fuzzy IF-THEN rules of the form

$$R^i: \quad \text{IF } x(k) \text{ is } X_{i1} \text{ AND ... AND } x(k-(n-1)) \text{ is } X_{in}$$
$$\text{THEN } y_i(k) = a_{i0} + a_{i1}x(k) + a_{i2}x(k-1) + ... + a_{in}x(k-(n-1)),$$

$$i=1,...,N,$$

with $x(k+1) = c_k y(k+1)$, $k = 0, 1, 2, ...$, in which

$$y(k+1) = \frac{\sum_{i=1}^{N} w_i \cdot y_i(k+1)}{\sum_{i=1}^{N} w_i}, \qquad k=0,1,2,....$$

Here, the fuzzy sets consist of intervals $\{X_j | j=1,...,n\}$ with the associate fuzzy membership functions $\{\mu_{X_j} | j=1,...,n\}$, and $\{w_i | i=1,...,N\}$ is a set of weights satisfying $w_i \geq 0$, $i=1,...,N$, and $\sum_{i=1}^{N} w_i > 0$.

In this definition, in all the rules R^i, $i=1,...,N$, X_{ij} share the same fuzzy subset X_j and the same membership function μ_{X_j}, for each $j=1,...,n$.

Note also that, similar to formula (4.19), the weights $\{w_i | i=1,...,N\}$ usually are chosen to be equal to $\mu_{Y_i}(y_i)$. Moreover, for simplicity of notation, in the following, let $c_k = 1$ for all $k=0,1,2,...$, and $a_{i0} = 0$ for all $i=1,...,N$, although this is not necessary in general.

Recall some concepts and stability results in classical systems theory.

Definition 4.2 A multi-input/multi-output (MIMO), nonlinear, discrete-time, dynamic system of the form

$$x(k+1) = f(x(k)), \quad x(k) \in R^m, \qquad k=0,1,2,\cdots,$$

is said to be *asymptotically stable* about an equilibrium point x_e, or x_e is an *asymptotically stable equilibrium point* of the system,

if

$$x_e = f(x_e)$$

and, starting from any $x(0) \in R^m$, all $x(k)$ are bounded and

$$x(k) \to x_e \qquad (k \to \infty).$$

In this definition, the convergence $x(k) \to x_e$ $(k \to \infty)$ is usually measured by the l_2-norm (the "length" of a vector), namely,

$$\lim_{k \to \infty} \| x(k) - x_e \|_2 = 0,$$

where $\| [x_1 \cdots x_m]^T \|_2 = \sqrt{x_1^2 + ... + x_m^2}$ is the length of a vector.

The following stability criterion is well known in classical systems theory.

Theorem 4.1 Suppose that the MIMO, nonlinear, discrete-time system

$$x(k+1) = f(x(k)), \quad x(k) \in R^m, \qquad k=0,1,2,...,$$

has an equilibrium point $x_e = 0$, and that there exists a scalar-valued function $V(x(k))$ satisfying

(i) $V(0) = 0$;

(ii) $V(x(k)) > 0$ for all $x(k) \neq 0$;

(iii) $V(x(k)) \to \infty$ as $\| x(k) \|_2 \to \infty$; and

(iv) $V(x(k+1)) - V(x(k)) < 0$ for all $x(k) \neq 0$, and all $k=0,1,2,\cdots$.

Then this system is asymptotically stable about the equilibrium point 0.

In this theorem, the function $V(x(k))$, if it exists, is called a *Lyapunov function*.

Note that this theorem particularly applies to all linear time-invariant state-space systems of the form

$$x(k+1) = A\, x(k), \quad x(k) \in R^m, \qquad k=0,1,2,\cdots. \quad (4.21)$$

For this linear time-invariant system, however, there is a very simple criterion for the determination of its asymptotic stability.

Theorem 4.2 Let λ_j, $j = 1, \ldots, m$, be eigenvalues, counting multiple ones, of the constant matrix A in the linear time-invariant system (4.21). Then the system is asymptotically stable about the equilibrium point 0 if and only if

$$|\lambda_j| < 1 \text{ for all } j = 1,\ldots,m.$$

Now, return to the single-input/single-output, linear, discrete-time dynamic fuzzy system described in Definition 4.1. To apply the classical stability Theorem 4.1 or 4.2 to it, one first reformulates it in the state-space setting as follows.

Let $c_k = 1$, $k = 0, 1, 2, \ldots$, and $a_{i0} = 0$, $i = 1, \ldots, N$, in the system (for simplicity, as mentioned above). Define

$$x(k) = [\, x(k)\ x(k-1)\ x(k-2)\ \ldots\ x(k-(n-1))\,]^\mathrm{T},$$

$$A_i = \begin{bmatrix} a_{i1} & a_{i2} & \cdots & a_{i,n-1} & a_{in} \\ 1 & 0 & \cdots & 0 & 0 \\ 0 & 1 & \cdots & 0 & 0 \\ \vdots & \vdots & \ddots & \vdots & 0 \\ 0 & 0 & \cdots & 1 & 0 \end{bmatrix}.$$

Although this (canonical) formulation has some redundancy (only the first equation is essential), it is convenient to use for stability analysis. Rewrite the system as

$$x(k+1) = \sum_{i=1}^{N} w_i A_i x(k) \bigg/ \sum_{i=1}^{N} w_i \qquad k=0,1,2,\cdots. \qquad (4.22)$$

Corollary 4.1 In the dynamic fuzzy system (4.22), let

$$A = \sum_{i=1}^{n} \frac{w_i}{\sum_{i=1}^{n} wi} A_i ,$$

and assume that $\{w_i\}$ are constants independent of k. Then, the system is asymptotically stable if and only if all eigenvalues of A, λ_i, $i=1,...,n$, satisfy

$$|\lambda_i| < 1, \qquad i=1,...,n.$$

Theorem 4.3 The discrete-time dynamic fuzzy system (3.22) is asymptotically stable about the equilibrium point 0 if there exists a common positive definite matrix P such that

$$A_i^T P A_i - P < 0, \qquad \text{for all } i=1,2,...,N.$$

D.2. Dynamic Fuzzy Systems with Control

Consider, once again, the typical single-input/single-output (SISO), linear, discrete-time, dynamic fuzzy system discussed in Definition 4.1.

Definition 4.3 An SISO, *discrete-time, dynamic fuzzy control system* is a fuzzy control model described by a set of fuzzy IF-THEN rules of the form

R^i: (IF $x(k)$ is X_{i1} AND ... AND $x(k-(n-1))$ is X_{in}) AND
        (IF $u(k)$ is U_{i1} AND ... AND $u(k-(m-1))$ is U_{im})
        THEN $y_i(k)$ $= a_{i0} + a_{i1}x(k) + ... + a_{in}x(k-(n-1))$
                 $+ b_{i0} + b_{i1}u(k) + ... + b_{im}u(k-(m-1)),$

     $i=1,...,N,$

with $m \le n$ and

$$x(k+1) = c_k \, y(k+1), \quad k=0,1,2,...,$$

in which

$$y(k+1) = \frac{\sum_{i=1}^{N} w_i y_i(k+1)}{\sum_{i=1}^{N} w_i} , \quad k=0,1,2,...,$$

where the fuzzy sets consist of intervals $\{X_j | j=1,...,n\}$ and $\{U_j | j=1,...,m\}$, as well as their associate fuzzy membership functions $\{\mu_{X_j} | j=1,...,n\}$ and $\{\mu_{U_j} | j=1,...,m\}$, respectively, and $\{w_i | i=1,...,N\}$ is a set of weights satisfying $w_i \geq 0$, $i = 1,...,N$, and $\sum_{i=1}^{N} w_i > 0$.

Here, note again that for each rule R^i, $i = 1,...,N$, all the X_{ij} share the same fuzzy subset X_j and the same membership function μ_{X_j}, for each $j = 1,...,n$. Also, the weights are usually quantified by the output membership functions in the same way as in (4.19). As before, for simplicity of notation and discussion, assume $c_k = 1$ for all $k = 0,1,2,...$, and $a_{i0} = b_{i0} = 0$ for all $i = 1,...,N$, in this model.

It is clear that Definitions 4.1 and 4.3 have no essential difference if the control inputs $\{u(k)\}$ in Definition 4.3 are independent of the states $\{x(k)\}$. In engineering control systems, however, most of the time one would like to have negative state-feedback controllers of the following form:

$$u(k) = -K_{11}x(k) - K_{12}x(k-1) - ... - K_{1n}x(k-(n-1)),$$

$$u(k-1) = -K_{22}x(k-1) - ... - K_{2n}x(k-(n-1)),$$

$$\vdots$$

$$u(k-m) = -K_{mm}x(k-m) - ... - K_{mn}x(k-(n-1)),$$

where K_{ij} are constant control gains to be determined.

To facilitate the discussion, consider a simple example. Suppose that a fuzzy control system is given by

$R_S^1 :$ IF $x(k)$ is X_1 AND $u(k)$ is U_1
            THEN $x_1(k+1) = a_1x(k) + b_1u(k)$.

$R_S^2 :$ IF $x(k)$ is X_2 AND $u(k)$ is U_2
            THEN $x_2(k+1) = a_2x(k) + b_2u(k)$.

Assume also that a state-feedback fuzzy controller is designed to be one described by

$R_C^1 :$ IF $x(k)$ is X_1 AND $x(k-1)$ is X_1
            THEN $u_1(k) = -K_{11}x(k) - K_{12}x(k-1)$.

$$R_C^2: \quad \text{IF } x(k) \text{ is } X_2 \text{ AND } x(k-1) \text{ is } X_2$$
$$\text{THEN } u_2(k) = -K_{21}x(k) - K_{22}x(k-1).$$

Then, one can combine all possibilities, where it is important to note that for "$x(k)$ is X_1 and $u(k)$ is U_1," there are two possibilities for the previous control input $u(k-1)$, either "is U_1" or "is U_2." And the same is true for "$x(k)$ is X_2 and $u(k)$ is U_1." One thus obtains

R^{11}:   (IF $x(k)$ is X_1) AND (IF $u(k)$ is U_1 AND $u(k-1)$ is U_1)

$$\text{THEN } x_{11}(k+1) \; := \; a_1x(k) + b_1u_1(k)$$
$$= (a_1 - b_1K_{11})\, x(k) - b_1K_{12}x(k-1).$$

R^{12}:   (IF $x(k)$ is X_1) AND (IF $u(k)$ is U_1 AND $u(k-1)$ is U_2)

$$\text{THEN } x_{12}(k+1) \; := \; a_1x(k) + b_1u_2(k)$$
$$= (a_1 - b_1K_{21})\, x(k) - b_1K_{22}x(k-1).$$

R^{21}:   (IF $x(k)$ is X_2) AND (IF $u(k)$ is U_2 AND $u(k-1)$ is U_1)

$$\text{THEN } x_{21}(k+1) \; := \; a_2x(k) + b_2u_1(k)$$
$$= (a_2 - b_2K_{11})\, x(k) - b_2K_{12}x(k-1).$$

R^{22}:   (IF $x(k)$ is X_2) AND (IF $u(k)$ is U_2 AND $u(k-1)$ is U_2)

$$\text{THEN } x_{22}(k+1) \; := \; a_2x(k) + b_2u_2(k)$$
$$= (a_2 - b_2K_{21})\, x(k) - b_2K_{22}x(k-1).$$

After all these state-feedback control inputs have been used, this fuzzy system, described by Definition 4.3, reduces to a new closed-loop dynamic fuzzy system, given by the rules R^{11}, R^{12}, R^{21}, and R^{22}, in which there are no more control variables $u(k)$. Thus, the stability conditions obtained in Theorem 4.3 can be applied, namely, the designer can choose the feedback control gains K_{11}, K_{12}, K_{21}, and K_{22} to ensure the asymptotic stability of the overall controlled system.

Example 4.6

Consider a fuzzy control system described by rule base

R_S^1:   IF $x(k)$ is X_1 AND $x(k-1)$ is X_1
         THEN $x_1(k+1) = 2.178x(k) - 0.588x(k-1) + 0.603u(k)$.

$R_S^2:$ IF $x(k)$ is X_2 AND $x(k-1)$ is X_2

         THEN $x_2(k+1) = 2.256x(k) - 0.361x(k-1) + 1.120u(k).$

Let the fuzzy state-feedback controller be described by

$R_C^1:$ IF $x(k)$ is X_1 AND $x(k-1)$ is X_1

         THEN $u_1(k) = r(k) - K_{11}x(k) - K_{12}x(k-1).$

$R_C^2:$ IF $x(k)$ is X_2 AND $x(k-1)$ is X_2

         THEN $u_2(k) = r(k) - K_{21}x(k) - K_{22}x(k-1).$

Here, $\{r(k)\}$ is a reference signal (set-points). Since the control input $u(k)$ does not appear in the condition parts, one may consider its membership values to be identically equal to 1 therein. The resulting closed-loop dynamic fuzzy system is then obtained as follows:

$R^{11}:$ IF $x(k)$ is X_1 AND $x(k-1)$ is X_1

         THEN $x_{11}(k+1) = (2.178-0.603K_{11})\,x(k)$
$$+ (-0.588-0.603K_{12})\,x(k-1)$$
$$+ 0.603\,r(k).$$

$R^{12}:$ IF $x(k)$ is X_1 AND $x(k-1)$ is X_2

         THEN $x_{12}(k+1) = (2.178-0.603K_{21})\,x(k)$
$$+ (-0.588-0.603K_{22})\,x(k-1)$$
$$+ 0.603\,r(k).$$

$R^{21}:$ IF $x(k)$ is X_2 AND $x(k-1)$ is X_1

         THEN $x_{21}(k+1) = (2.256-1.120K_{11})\,x(k)$
$$+ (-0.361-1.120K_{12})\,x(k-1)$$
$$+ 1.120\,r(k).$$

$R^{22}:$ IF $x(k)$ is X_2 AND $x(k-1)$ is X_2

         THEN $x_{22}(k+1) = (2.256-1.120K_{21})\,x(k)$
$$+ (-0.361-1.120K_{22})\,x(k-1)$$
$$+ 1.120\,r(k).$$

The overall state output $x(k+1)$ is then computed by

$$x(k+1) = \frac{w_1 x_{11}(k+1) + w_2 x_{12}(k+1) + w_3 x_{21}(k+1) + w_4 x_{22}(k+1)}{w_1 + w_2 + w_3 + w_4}$$

for some weights $w_i \geq 0$, $i=1,2,3,4$, and $\sum_{i=1}^{4} w_i > 0$.

Suppose now that the membership functions μ_{X_1} and μ_{X_2} are given as shown in Figure 4.8.

Then, the weights can be determined by using these membership functions, as discussed before. For simplicity, assume that the reference signal $r(k) = 0$ for all $k = 0,1,2,...$, and that the controller in both R_C^1 and R_C^2 are the same:

$$u_1(k) = u_2(k) = -K\, x(k), \qquad k=0,1,2,....$$

Then, the closed-loop fuzzy system reduces to

R^1: IF $x(k)$ is X_1 AND $x(k-1)$ is X_1
    THEN $x_1(k+1) = (2.178-0.603K)\, x(k) - 0.588\, x(k-1)$.

R^2: IF $x(k)$ is X_2 AND $x(k-1)$ is X_2
    THEN $x_2(k+1) = (2.256-1.120K)\, x(k) - 0.361\, x(k-1)$.

Finally, by applying the above stability theorems, one can verify that this system is stable.

Problems

P4.1 Write a rule base for the following statement, and show the membership functions that you chose for all your descriptions:

"The output is positive large, unless two inputs have one negative small and one positive small, or unless they have one negative medium or one positive medium, and in these two exceptional cases the output is positive medium."

P4.2 Write a rule base for the following logical statement, using as few as possible numbers of rules:

"If a and b are both real numbers, then let $y = 0$, unless $a < b$ (for which $y = a + b$) or unless $a > b$ (for which $y = ab$); if either a or b is a complex number then let $y = \infty$."

P4.3 Compute the following logical operations:
(a) IF $\mu(a) = 0.9$ AND $\mu(a \Rightarrow b) = 0.2$ THEN $\mu(b) = ?$
(b) IF $\mu(\overline{a}) = 0.8$ AND $\mu(a \Rightarrow b) = 0.8$ THEN $\mu(b) = ?$

P4.4 Show all steps in obtaining the three output membership functions $\mu_S(y)$, $\mu_{L1}(y)$, and $\mu_{L2}(y)$ in Example 4.4 of the Lecture Notes.

P4.5 Can the following statements be converted to be fuzzy rules and, if so, how?
(a) "IF a is A THEN b is not B AND not c is C"
(b) "IF a is A OR b is B THEN c is $\overline{C}$"
(c) "IF a is A UNLESS b is not B OR c is C"

CHAPTER 5

Fuzzy Control Systems

Control systems theory can be traced back to the age of World War II, or even earlier, when the design, analysis, and synthesis of servomechanisms were essential in the manufacturing of electromechanical systems. The development of control systems theory has since gone through an evolutionary process, starting from some basic, simplistic, frequency-domain analysis for single-input single-output (SISO) linear control systems, to a mathematically sophisticated modern theory of multi-input multi-output (MIMO) linear and nonlinear systems described by differential and/or difference equations.

It is believed that the advances of space technology in the 1950s completely changed the spirit and orientation of the classical control systems theory: the challenges posed by the high accuracy and extreme complexity of the space systems, such as space vehicles and structures, stimulated and promoted the existing control theory very strongly, developing it to such a high mathematical level that can use many new concepts like state-space and optimal controls. The theory is still rapidly growing today; it employs advanced mathematics such as differential geometry, operator theory, and functional analysis, and connects with many theoretical as well as applied sciences like artificial intelligence, soft computing, and the Internet technology.

The existing modern control systems theory, referred to as *conventional* or *classical* control systems theory in this text, has been extensively developed. The theory is now relatively complete for linear control systems, and has taken the lead in modern technology and industrial applications where control and automation are fundamental. It generally can provide rigorous analysis and even perfect solutions when a system is defined in precise mathematical terms. In addition to these advances, adaptive and robust as well as nonlinear systems control theories have also seen very rapid development in the last two decades.

Conventional mathematics and control theory exclude vagueness and incomplete, particularly contradictory conditions. As a consequence, conventional control systems theory does not attempt to study any formulation, analysis, and control of what has been called *fuzzy systems*, which may be vague, incomplete, linguistically described, or even inconsistent, and yet are ubiquitous in the real world.

Fuzzy set theory and fuzzy logic, studied in some detail previously, play a central role in the investigation of controlling such systems. The main contribution of *fuzzy control theory*, as a new alternative and branch of control systems theory that uses fuzzy logic, is its ability to handle many practical problems that cannot be adequately managed by conventional control theories and techniques. At the same time, the results of fuzzy control theory are consistent with the existing classical ones when the system under control goes from being fuzzy to non-fuzzy.

The fuzzy control systems theory is developed for solving real-world problems. The real-world problems exist in the first place. Fuzzy logic, fuzzy set theory, fuzzy modeling, fuzzy control methods, etc. are all man-made and subjectively introduced to the scene. If this fuzzy interpretation is correct and if the fuzzy mathematical theory works, then one should be able to solve the real-world problems after the fuzzy operations have been completed in the fuzzy environment and then the entire process is finally returned to the original real-world setting. This is what is called the "fuzzification – fuzzy operation – defuzzification" routine in the fuzzy control systems theory.

Although fuzzy control systems theory is used for handling imprecisely given systems, this theory itself is not vague. On the contrary, it is rigorous: as has been seen from Chapters $1-3$, fuzzy logic is a certain type of logic, and fuzzy mathematics is a branch of mathematics. It will be seen in this chapter that fuzzy logic control is an effective logical control technique using some simple fuzzy logic, in a way comparable to the well-known programmable logic control (PLC), which uses the classical two-valued logic.

To motivate the study, the PLC theory is first reviewed in the next section as an introduction to the general fuzzy logic control (FLC) theory and methodology.

I. CLASSICAL PROGRAMMABLE LOGIC CONTROL

A programmable logic controller (PLC) is a microprocessor-based, specialized computer that carries out logical control functions of many types. Its general structure is shown in Figure 5.1.

PLC usually operates on ordinary house current but may control circuits of large amperage and voltage of 440V and higher. A PLC typically has a detachable *programmer* module, used to construct a new program or modify an existing one, which can be taken easily from one PLC to another. The program instructions are typed into the memory by the user using a keyboard. The central processing unit (CPU) is the heart of the PLC, which has three parts: the processor, memory, and power supply. In the input-output (I/O) modules, the input module has terminals into which the user enters outside process electrical signals and the output module has another set of terminals that send action signals to the processor. A remote electronic system for

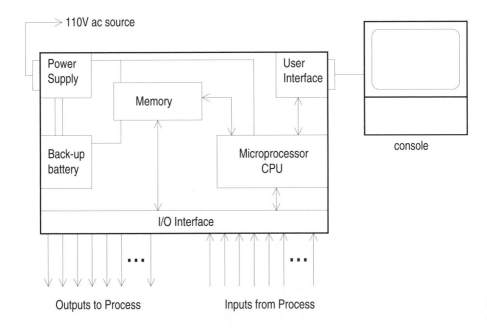

Figure 5.1   A programmable logic controller (PLC)

connecting the I/O modules to long-distance locations can be added, so that the actual operating process under the PLC control can be some miles away from the CPU and its I/O modules.   Optional units in a PLC include racks, chassis, printer, program recorder/player, etc.

To see how a PLC works, consider an example of a simple pick-and-place robot arm working in an automatic cycle under the control of a PLC.   This simple robot arm is shown in Figure 5.2, which has three joints: one rotational type and two sliding types, as indicated by the arrows; and has a gripper that can close (to grasp an object) and open (to release the object).

To control this robot arm, according to the structure shown in Figure 5.2, one needs eight actions created by the PLC, called *outputs* of the PLC, as follows:

       A0:   rotate the base counterclockwise (CCW),

       A1:   rotate the base clockwise (CW),

       A2:   lift the arm,

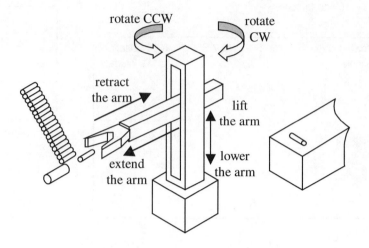

Figure 5.2 A pick-and-place robot arm controlled by a PLC

    A3:   lower the arm,

    A4:   extend the arm,

    A5:   retract the arm,

    A6:   close the gripper,

    A7:   open the gripper.

Suppose that in each cycle of the control process, the entire job performance is completed in 12 steps sequentially within 40 seconds, as shown in Figure 5.3. In this figure, it is shown that the PLC is needed to first operate the "extend arm" action for 4 seconds (from the beginning), and then stop this action. To this end, the PLC operates the "lower arm" action for 2 seconds and then stops, and so on. For simplicity, assume that the transitions between any two consecutive steps are continuous and smooth, so that no time is needed for any of these transitions.

To set up the PLC timer, called the *drum-timer*, so as to accomplish these sequential activities of the robot arm, one needs to select the time frequency for the PLC. Since the smallest step time in Figure 5.3 is the 1-second gripper operation, one selects a count frequency of 1 count per second. The entire PLC drum-timer array is shown in Table 5.1, where 1 means ON and 0 means OFF, as usual.

Table 5.1 PLC Drum-Timer Array for the Robot Arm

Step	Counts	Outputs							
	(count/sec)	A0	A1	A2	A3	A4	A5	A6	A7
1	4	0	0	0	0	1	0	0	0
2	2	0	0	0	1	0	0	0	0
3	1	0	0	0	0	0	0	1	0
4	2	0	0	1	0	0	0	0	0
5	4	0	0	0	0	0	1	0	0
6	7	1	0	0	0	0	0	0	0
7	4	0	0	0	0	1	0	0	0
8	2	0	0	0	1	0	0	0	0
9	1	0	0	0	0	0	0	0	1
10	2	0	0	1	0	0	0	0	0
11	4	0	0	0	0	0	1	0	0
12	7	0	1	0	0	0	0	0	0

The above-described sequential control process is by nature an open-loop control procedure. Now, if one uses sensors to provide *feedback* information at each step of the arm's motion, then a closed-loop logic control can also be performed. For this purpose, mechanical limit switches (or proximity switches) can be used to sense the point of completion of each arm motion and, hence, provide *inputs* to the PLC. The PLC can then be programmed not to begin the next step until the previous step has been completed, regardless of how much time this might take.

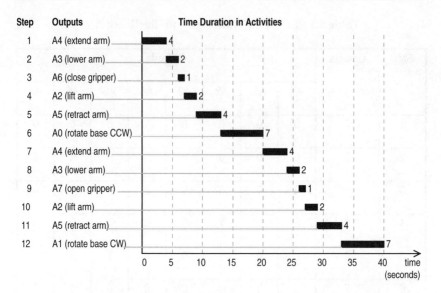

Figure 5.3 Time duration of activities of the robot arm

By industrial convention, the number symbol 00 is used to represent the input from the limit switch that signals completion of the previous output A0 ("rotating the base counterclockwise"), the number symbol 01 to represent the input from the limit switch that signals completion of the previous output A1 ("rotating the base clockwise"), and so on, and finally 07 for the input that signals completion of A7. The process is illustrated by the 12 steps shown in Figure 5.4.

The entire control scheme, together with necessary prior inputs for each arm motion, is shown in Table 5.2. Note that in this diagram, three previous steps are given to be safe and sufficient as prerequisite; however, it is unnecessary to do so in a real design, since for many of them two previous steps are enough for distinction of different actions.

Table 5.2 Robot Arm Control Scheme

Arm Motion	PLC Outputs	PLC Inputs	Necessary Prior Conditions
extend arm	A4	04	02, 05, 01
lower arm	A3	03	05, 01, 04
close gripper	A6	06	01, 04, 03
lift arm	A2	02	04, 03, 06
retract arm	A5	05	03, 06, 02
rotate base CCW	A0	00	06, 02, 05
extend arm	A4	04	02, 05, 00
lower arm	A3	03	05, 00, 04
open gripper	A7	07	00, 04, 03
lift arm	A2	02	04, 03, 07
retract arm	A5	05	03, 07, 02
rotate base CW	A1	01	07, 02, 05

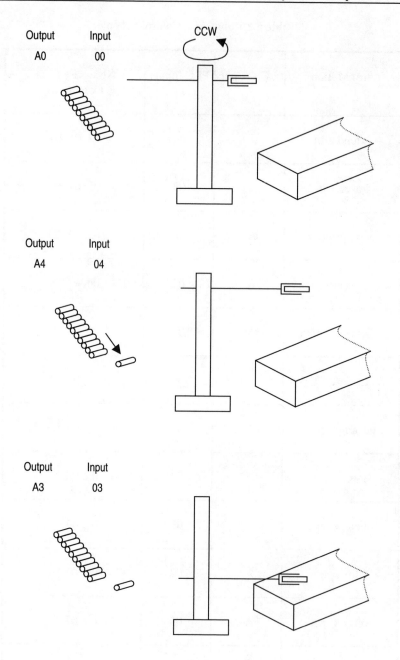

Output   Input
A0   00

CCW

Output   Input
A4   04

Output   Input
A3   03

Output   Input
  A7   07

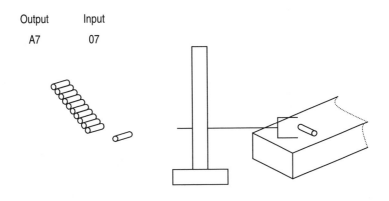

Output   Input
  A2   02

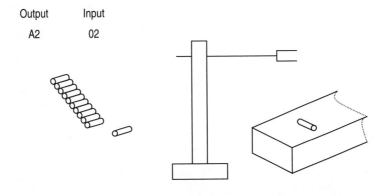

Output   Input
  A5   05

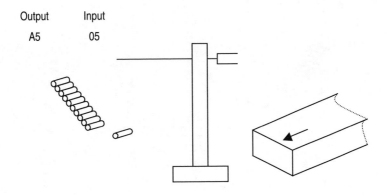

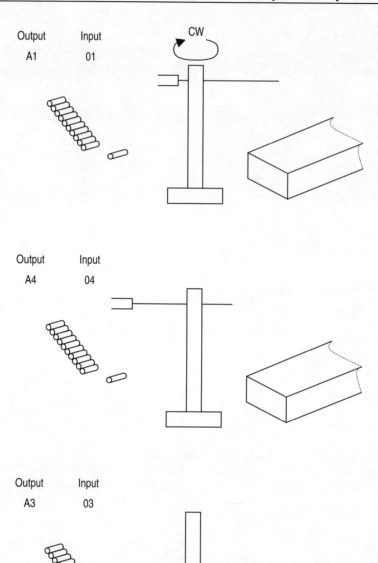

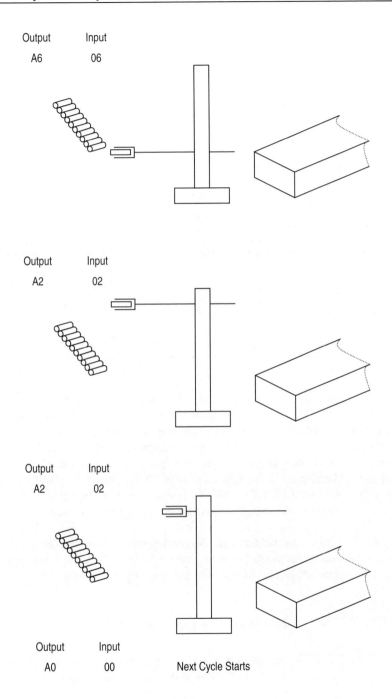

Figure 5.4 The control process of the robot arm by PLC

Table 5.3 Ladder Diagram Symbols

─┤├─	Input contacts	
─○─	Output loads	
A ─┤╱├─	Logical NOT	$\overline{A}$
A B ─┤├─┤├─	Logical AND	$A \wedge B$
A ─┤├─┐ ─┤├─┘ B	Logical OR	$A \vee B$

It is also important to remark that if one does not consider the necessary prior conditions, then when the PLC applies the logic control to the robot arm, it may make mistakes. For example, the two steps "rotate base CCW" and "rotate base CW" in Table 5.2 both have the previous steps 02 and then 05. Hence, it is necessary to provide one more previous step to one of them, for instance, indicating there is a step 07 prior to "Rotate base CW" (which is not prior to "rotate base CCW"). This can distinguish these two steps, so as to ensure the program take correct actions for the two different situations.

Figure 5.5 shows the ladder logic diagram corresponding to Table 5.2, in which the standard symbols listed in Table 5.3 are used. Also, in this figure, the input contact P0 is used to represent a power-ON switch, and

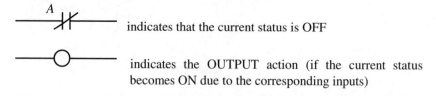

─┤╱├─   indicates that the current status is OFF

─○─   indicates the OUTPUT action (if the current status becomes ON due to the corresponding inputs)

Here, the input contacts include switches, relays, photoelectric sensors, limit switches, and other ON-OFF signals from the logical system; the output loads

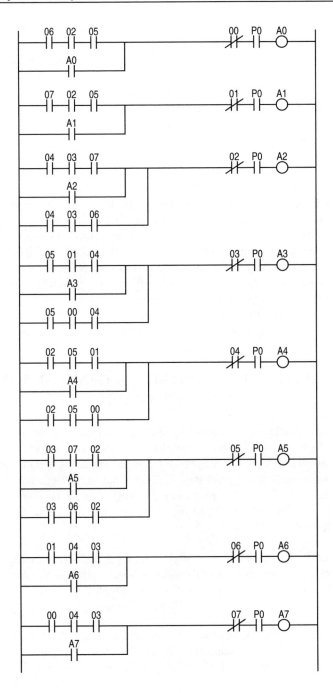

Figure 5.5 PLC ladder logic diagram for the robot arm

include motors, valves, alarms, bells, lights, actuators, or other electrical loads to be driven by the logical system. One should not confuse the input contact symbol with the familiar circuit symbol for capacitors. Also, one should note that output loads can become input contacts in the steps to follow, depending on the actual system structure. Finally, it should be remarked that a control process described by Figure 5.5 and Table 5.2 is not the only option; one may change the prior conditions provided that these conditions can distinguish different stages of the robot arm motion and are sufficient to determine all the control actions.

In a routine yet somewhat tedious procedure, one can verify that the PLC ladder logic diagram shown in Figure 5.4 performs the automatic control process summarized in Table 5.2. Under the indicated necessary prior conditions for each step throughout the process, this PLC works for the designed "pick-and-place" motion-control of the robot arm described by Figure 5.3. The design procedure is time-consuming. However, once a design has been completed, the PLC can be automatically programmed over and over again, to control the robot arm to repeat the same "pick-and-place" motion for the production line. When the job assignment is changed, the PLC has the flexibility to be reprogrammed to perform a new logic control process.

II. FUZZY LOGIC CONTROL: A GENERAL MODEL-FREE APPROACH

The programmable logic controllers (PLCs) discussed in the last section, which have been widely used in industries, can only perform classical (ON-OFF) two-valued logic in programming some relatively simple but very precise automatic control processes. To carry out fuzzy logic-based control programming, a new type of programmable fuzzy logic controllers is needed that can take infinitely many values between 1 (completely ON) and 0 (completely OFF).

The majority of fuzzy logic control systems are *knowledge-based systems* in that either their fuzzy models or their fuzzy logic controllers are described by fuzzy IF-THEN rules, which have to be established based on experts' knowledge about the systems, controllers, performance, etc.

In fuzzy control, the introduction of input-output intervals and membership functions is more or less subjective, depending on the designer's experience and knowledge about the plant, and the available information. However, it should be emphasized that after the determination of the fuzzy sets, all mathematics to follow are rigorous. Also, the purpose of designing and applying fuzzy logic control systems is, above all, to tackle those vague, ill-described, and complex plants and processes that can hardly be handled by

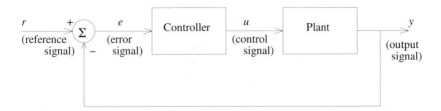

Figure 5.6 A typical closed-loop set-point tracking system

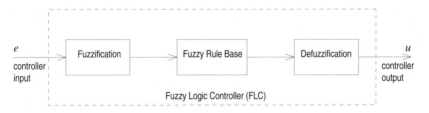

Figure 5.7 General structure of a fuzzy logic controller

classical systems theory, classical control techniques, and classical two-valued logic.

The first type of fuzzy logic control system is that the fuzzy logic controller directly performs the control actions and thus completely replaces a conventional control algorithm. Yet, there is another type of fuzzy logic control system: the fuzzy logic controller is involved in a conventional control system and thus becomes part of the mixed control algorithm, so far as to carry out or improve the performance of the overall control system.

In this section, a general approach of fuzzy logic control for a conventional system (plant or process) of the feedback type is introduced.

A. Closed-Loop Set-Point Tracking System

Consider the typical continuous-time, closed-loop, set-point tracking system shown in Figure 5.6.

In this figure, assume that the plant is a conventional (crisp) one, which is given but its mathematical model may not be known, and suppose that all the signals (r, e, and y) are crisp. The closed-loop set-point tracking control problem is to design the controller such that the output signal of the controlled plant, y, can track the given reference signal r (called a set-point; but need not be a constant):

$$e(t) := r(t) - y(t) \to 0 \qquad (t \to \infty).$$

Instead of designing a conventional controller, in the following a fuzzy logic controller is designed for the same purpose. Recall that in designing a conventional controller, a precise mathematical model (formulation) of the plant is usually necessary. Here, however, to design a fuzzy logic controller, the mathematical formulation of the plant is supposed to be completely unknown (a "black box").

If the mathematical formula of the plant is unknown, how can one design a controller to perform the required set-point tracking? One may think of designing a conventional controller and try to answer the question at this point, to appreciate the ease of fuzzy logic controller design to be studied below.

First, some concepts and terminologies are introduced. The general structure of a *fuzzy logic controller* (FLC), or *fuzzy controller* (FC) for short, consists of three basic portions: (i) the *fuzzification* unit at the input terminal, (ii) the *inference engine* built on the fuzzy *logic control rule base* in the core, and (iii) the *defuzzification* unit at the output terminal, see Figure 5.7.

The fuzzification module transforms the physical values of the current process signal, the error signal in Figure 5.6, which is input to the fuzzy logic controller, into a normalized fuzzy set consisting of a set (interval) for the range of the input values and an associate membership function describing the properties or degrees of the confidence of the input belonging to this range.

The role of the inference engine in the FLC is key to making the controller work and work effectively. A typical fuzzy logic IF-THEN rule base performing the inference is a set of IF-THEN rules of the form:

R^1:   IF input e_1 is E_{11} AND ... AND input e_n is E_{1n}
            THEN output u_1 is U_1.

$\vdots$

R^m: IF input e_1 is E_{m1} AND ... AND input e_n is E_{mn}
            THEN output u_m is U_m.

Here, the fuzzy sets E_{11}, ..., E_{m1} share the same set (interval) E_1 and the same membership function μ_{E_1} defined on E_1, and fuzzy sets (intervals) E_{1n}, ..., E_{mn} share the same set (interval) E_n and the same membership function μ_{E_n} defined on E_n. In general, m rules produce m controller outputs, u_1, ..., u_m, belonging to m fuzzy subsets (intervals), U_1, ..., U_m, in which, of course, some of them may overlap.

The establishment of this rule base depends heavily on the designer's work experience, knowledge about the physical plant, analysis and design skills, etc., and is, hence, more or less subjective. A good design can make the controller work; a better design can make it work more effectively. Yet, there are some general criteria and some routine steps for the designer to follow in a real design, which will be discussed in more detail later.

The defuzzification module is the connection between the control rule base and the physical plant to be controlled, which plays the role of a transformer mapping the controller outputs (generated by the control rule base in fuzzy terms) back to the crisp values that the plant can accept.

The controller outputs u_1, ..., u_m, generated by the rule base, are fuzzy signals belonging to the fuzzy sets U_1, ..., U_m, respectively. The defuzzification module converts these fuzzy controller outputs to a pointwise and crisp real signal, u, and then send it to the physical plant as a control action for tracking.

This tree-step design routine, the "fuzzification – rule base establishment – defuzzification" procedure, is further discussed in detail next.

B. Design Principle of Fuzzy Logic Controllers

Suppose that a scalar-valued constant reference signal r, the set-point, is given, which needs not be a constant in general. A plant is assumed given, however, with an unknown mathematical formulation. The aim is to design a fuzzy logic controller, to be put into the controller block of Figure 5.6, to drive the plant output $y(t)$ to track the constant set-point r after long-enough time, namely,

$$e(t) := r - y(t) \to 0 \qquad (t \to \infty).$$

The following is a general approach for the design of the fuzzy logic controller.

(1) The Fuzzification Module

(1.1) This module transforms the physical values (position, voltage, degree, etc.) of the process signal, the error signal shown in Figure 5.7 as an input to the fuzzy logic controller, into a normalized fuzzy set consisting of a set (interval) for the range of the input values and a normalized membership function describing the properties (e.g., large or small) or degree of confidence of the input belonging to this range.

For example, suppose that the physical error signal is temperature, and its distribution is within the range $[-25°, 45°]$ in an application in which the fuzzy logic controller needs to be designed to reduce this error to $0° \pm 0.01°$. In this

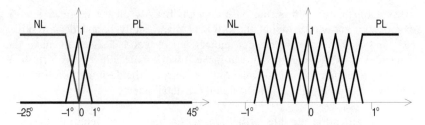

Figure 5.8 Selection of membership functions for error signals

case, the scale of 1° is too large to use in the measurement of the control effect within ±0.01°. Hence, one may first rescale the range to be [−2500,4500] in the unit of 0.01°. But this is not convenient in the calculation of fuzzy membership functions either. To have a compromise, a better choice can be a combination of two different scales (see Figure 5.8):

Use [−25°,45°] when |error| > 1;

Use [−100*,100*] when |error| ≤ 1, with 1° = 100*.

(1.2) This module selects reasonable and good, ideally optimal, membership functions under certain convenient criteria, which are meaningful to the application at hand.

In the above-described temperature control example, one may use the membership functions shown in Figure 5.8 to describe "the error is positive large (PL)," or "the error is negative large (NL)," etc., in which one figure is in the 1° scale and the other is in the 1* = 0.01° scale. A designer may have other choices, of course, as long as it is reasonable for the application in consideration. But this example should provide some idea about how the selection can be done in general.

In the fuzzification module, the input is a crisp physical signal (e.g., temperature) of the real process and the output is a fuzzy set consisting of intervals and membership functions. This output will be the input to the next module, the fuzzy logic IF-THEN rule base for the control, which requires fuzzy-set inputs in order to be compatible with the fuzzy logic rules.

(2) The Fuzzy Logic Rule Base

Designing a good fuzzy logic rule base is a key to obtaining a satisfactory controller for a particular application. Classical analysis and control strategies should be incorporated in the establishment of a rule base. A general procedure in designing a fuzzy logic rule base includes the following:

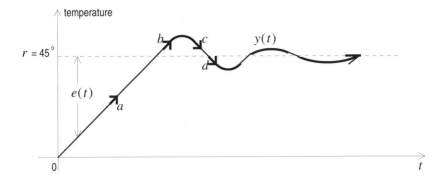

Figure 5.9 Temperature set-point tracking example

(2.1) Determining the process states and control variables

In the set-point tracking example discussed above, suppose that the set-point r is a target temperature to be reached, say $r = 45°$. The process state is the controlled system output, $y(t)$, which is also temperature. Finally, the error signal

$$e(t) = r - y(t), \tag{5.1}$$

as shown in Figure 5.9, is used to create the control variable u (see Figure 5.7) through the controller. Here, assume $y(0) = 0$ for simplicity.

(2.2) Determining input variables to the controller

The tracking error signal $e(t)$ is an input variable to the controller. Oftentimes, one needs more auxiliary input variables in order to establish a complete and effective rule base.

In this temperature control example, it can be easily seen that only the error signal $e(t)$ is not enough to write an IF-THEN control rule. Indeed, suppose $e > 0$ at a moment. Then

$$e = r - y > 0 \qquad \text{or} \qquad r > y;$$

namely, at that moment the system output y is below the set-point. This indicates that the output y is either at position a or position d in Figure 5.9. However, this information is not sufficient for determining a control strategy that can bring the trajectory of y to approach the set-point thereafter: if the output y is at position a then the controller should take action to keep the trajectory going up; if y is at position d then the controller should turn the trajectory to the opposite moving direction (from pointing down to pointing up). Therefore, one more input variable that can distinguish these two situations is necessary. Same discussions apply to points b and c.

Recall from Calculus that if a curve is moving up then its derivative is positive and if it is moving down then its derivative is negative. Hence, the change of the error signal, denoted $\dot{e}$ or Δe, can help distinguish the two situations at points a and d, as well as that at points b and c, shown in Figure 5.9. One may introduce the change of error as the second input variable for this temperature set-point tracking example.

More input variables can be introduced such as the second derivative $\ddot{e}$ or the sum (integral) Σe of the errors, to make the controller work more effectively. However, this will also complicate the design, and so there is a trade-off for the design engineer. In order to simplify the design and for simplicity of discussion, in this example only two input variables, e and $\dot{e}$, are used for the controller.

(2.3) Establishing a fuzzy logic IF-THEN rule base

In Figure 5.9, it is clear that essentially one has four situations to consider: when the temperature output $y(t)$ is at the situations represented by the points a, b, c, and d. One thus needs at least four rules for this set-point tracking application in order to make it work and work effectively. One simple choice is:

R^1:   IF $e > 0$ AND $\dot{e} < 0$ THEN $u(t+) = C \cdot u(t)$;

R^2:   IF $e < 0$ AND $\dot{e} < 0$ THEN $u(t+) = -C \cdot u(t)$;

R^3:   IF $e < 0$ AND $\dot{e} > 0$ THEN $u(t+) = C \cdot u(t)$;

R^4:   IF $e > 0$ AND $\dot{e} > 0$ THEN $u(t+) = -C \cdot u(t)$.

Otherwise (e.g., $e = 0$ or $\dot{e} = 0$), $u(t+) = u(t)$, till next step.

Here, C is a positive constant usually determined by tuning. As will be seen below, this simple design does work, though may not be the most effective.

In Figure 5.9, it is very important to observe that

$$\dot{e}(t) = \dot{r} - \dot{y}(t) = -\dot{y}(t). \tag{5.2}$$

Rules R^1, R^2, R^3, and R^4 correspond to the situations indicated by points a, b, c, and d, respectively, in Figure 5.9. In these rules, the notation $u(t+) = C \cdot u(t)$ means that the controller continues its previous action unchanged, with some re-scaling only, while $u(t+) = -C \cdot u(t)$ means that the controller corrects its previous action by replacing it with a re-scaled opposite action. Quantitative implementation of these actions will be further discussed below.

To this end, it can be easily realized that by following the above four rules, the controller is able to drive the temperature output $y(t)$ to track the set-point r, at least in principle. For example, if $e > 0$ and $\dot{e} < 0$ then $r > y$ and $\dot{y} > 0$, which means that the curve y is at position a of Figure 5.9. In this case, rule R^1 implies that the controller should maintain its current action (to keep pushing the system in the same way), with some rescaling of the strength. The other three rules can be similarly interpreted.

Some technical issues remain. Consider the situation at point a in Figure 5.9, which fires control rule R^1. One problem with this rule is that if position a is far away from the set-point, this rule would let the controller keep driving the output temperature up toward the set-point, but very slowly; so it may need a long time to complete the task. Conversely, if position a is close to the set-point, this rule would lead the controller to drive the output temperature overshooting the set-point. As a result of such rules and actions, the output temperature will oscillate up and down around the set-point, which may not eventually settle at the set-point, or take a very long time to do so. Hence, these four rules have to be further improved to be useful.

The basic idea for improving the control performance is to use membership values to weight the control actions. More precise description is given below.

For a simple discussion, only simple membership functions are used, as shown in Figure 5.10, rather than those shown in Figure 5.8, for both e and $\dot{e}$. In a real application, of course, a designer can (and likely should) employ more membership functions.

Using these membership functions as weights for the control $u(t)$, one expects to accomplish the following task: if y is far away from r then the control action is large, but if y is close to r then the control action is small. The improved control rule base is obtained as follows:

R^1:   IF e = PL AND $\dot{e} < 0$ THEN $u(t^+) = \mu_{PL}(e) \cdot u(t)$;

R^2:   IF e = PS AND $\dot{e} < 0$ THEN $u(t^+) = (1-\mu_{PS}(e)) \cdot u(t)$;

R^3:   IF e = NL AND $\dot{e} < 0$ THEN $u(t^+) = -\mu_{NL}(e) \cdot u(t)$;

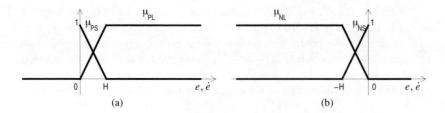

Figure 5.10 Four membership functions for both e and $\dot{e}$

R^4:   IF $e = $ NS AND $\dot{e} < 0$ THEN $u(t^+) = -(1-\mu_{NS}(e)) \cdot u(t)$;

R^5:   IF $e = $ NL AND $\dot{e} > 0$ THEN $u(t^+) = \mu_{NL}(e) \cdot u(t)$;

R^6:   IF $e = $ NS AND $\dot{e} > 0$ THEN $u(t^+) = (1-\mu_{NS}(e)) \cdot u(t)$;

R^7:   IF $e = $ PL AND $\dot{e} > 0$ THEN $u(t^+) = -\mu_{PL}(e) \cdot u(t)$;

R^8:   IF $e = $ PS AND $\dot{e} > 0$ THEN $u(t+1) = -(1-\mu_{PS}(e)) \cdot u(t)$.

Otherwise, $u(t+1) = u(t)$.

Here and below, "= PL" means "is PL," etc.

To implement these rules on a digital computer, one actually uses their discrete-time version of the form

$$u((k+1)T) = u(kT) + \Delta u(kT) \qquad (5.3)$$

so that

$$u(t^+) = u((k+1)T) \quad \text{and} \quad u(t) = u(kT),$$

where T is the sampling time and $\Delta u(kT)$ describes the new change of the control action at $t = kT$, $k = 0, 1, 2, \cdots$. Thus, the actual executive rules on a computer program become the following:

Executive Rule Base

R^1:   IF $e(kT) = $ PL AND $\dot{e}(kT) < 0$
          THEN $u((k+1)T) = \mu_{PL}(e(kT)) \cdot u(kT)$;

R^2:   IF $e(kT) = PS$ AND $\dot{e}(kT) < 0$
       THEN $u((k+1)T) = (1-\mu_{PS}(e(kT))) \cdot u(kT)$;

R^3:   IF $e(kT) = NL$ AND $\dot{e}(kT) < 0$
       THEN $u((k+1)T) = -\mu_{NL}(e(kT)) \cdot u(kT)$;

R^4:   IF $e(kT) = NS$ AND $\dot{e}(kT) < 0$
       THEN $u((k+1)T) = -(1-\mu_{NS}(e(kT))) \cdot u(kT)$;

R^5:   IF $e(kT) = NL$ AND $\dot{e}(kT) > 0$
       THEN $u((k+1)T) = \mu_{NL}(e(kT)) \cdot u(kT)$;

R^6:   IF $e(kT) = NS$ AND $\dot{e}(kT) > 0$
       THEN $u((k+1)T) = (1-\mu_{NS}(e(kT))) \cdot u(kT)$;

R^7:   IF $e(kT) = PL$ AND $\dot{e}(kT) > 0$
       THEN $u((k+1)T) = -\mu_{PL}(e(kT)) \cdot u(kT)$;

R^8:   IF $e(kT) = PS$ AND $\dot{e}(kT) > 0$
       THEN $u((k+1)T) = -(1-\mu_{PS}(e(kT))) \cdot u(kT)$,

Otherwise, $u((k+1)T) = u(kT)$; for all $k = 0, 1, 2, \cdots$, where $\dot{e}(kT) \approx \frac{1}{T}[e(kT) - e((k-1)T)]$, with the initial conditions

$$y(0) = 0, \ e(-T) = e(0) = r - y(0), \ \dot{e}(0) = \frac{1}{T}[e(0) - e(-T)] = 0.$$

Here, of course, the simple treatment "Otherwise, $u((k+1)T) = u(kT)$; for all $k = 0, 1, 2, \cdots$" may be further modified to obtain a more effective (though slightly more complicated) rule base, as seen from the following simulation examples.

At this point, it is illustrative to show a simple simulation example, to verify that the above-described simple-minded idea and natural design of the rule base indeed work.

In this simulation example of temperature control, the input membership functions shown in Figure 5.10 are used, together with constant control inputs for simplicity.

Introduce the following convenient notation:

C1 – constant control input for Rules 1, 3, 5, 7, when errors are outside the marked point H or $-H$ shown in Figure 5.10.

C2 – constant control input for Rules 2, 4, 6, 8, when errors are inside the marked point H or $-H$ shown in Figure 5.10.

The complete Executive Rule Base is as follows:

R^1:   IF $e(kT) = $ PL AND $\dot{e}(kT) < 0$
         THEN $u((k+1)T) = \mu_{PL}(e(kT)) \cdot$ C1;

R^2:   IF $e(kT) = $ PS AND $\dot{e}(kT) < 0$
         THEN $u((k+1)T) = (1-\mu_{PS}(e(kT))) \cdot$ C2;

R^3:   IF $e(kT) = $ NL AND $\dot{e}(kT) < 0$
         THEN $u((k+1)T) = -\mu_{NL}(e(kT)) \cdot$ C1;

R^4:   IF $e(kT) = $ NS AND $\dot{e}(kT) < 0$
         THEN $u((k+1)T) = -(1-\mu_{NS}(e(kT))) \cdot$ C2;

R^5:   IF $e(kT) = $ NL AND $\dot{e}(kT) > 0$
         THEN $u((k+1)T) = \mu_{NL}(e(kT)) \cdot (-$ C1$)$;

R^6:   IF $e(kT) = $ NS AND $\dot{e}(kT) > 0$
         THEN $u((k+1)T) = (1-\mu_{NS}(e(kT))) \cdot (-$ C2$)$;

R^7:   IF $e(kT) = $ PL AND $\dot{e}(kT) > 0$
         THEN $u((k+1)T) = -\mu_{PL}(e(kT)) \cdot$ C1;

R^8:   IF $e(kT) = $ PS AND $\dot{e}(kT) > 0$
         THEN $u((k+1)T) = -(1-\mu_{PS}(e(kT))) \cdot$ C2,

R^9:   IF $e(kT) = $ (NL or NS) AND $\dot{e}(kT) = 0$
         THEN $u((k+1)T) = \mu_{NL}(e(kT)) \cdot$ C1 $+ \mu_{NS}(e(kT)) \cdot$ C2 ;

R^{10}:   IF $e(kT) = $ (PL or PS) AND $\dot{e}(kT) = 0$
         THEN $u((k+1)T) = \mu_{PL}(e(kT)) \cdot$ C1 $+ \mu_{PS}(e(kT)) \cdot$ C2 ;

Here, C1 > C2 > 0 and R^9 and R^{10} are used to implement the "Otherwise" part of the original rule base.

A typical temperature tracking simulation result is shown in Figure 5.11, where the set-point is 20°, $T = 1$, $H = 5$, C1 = 5, and C2 = 2.

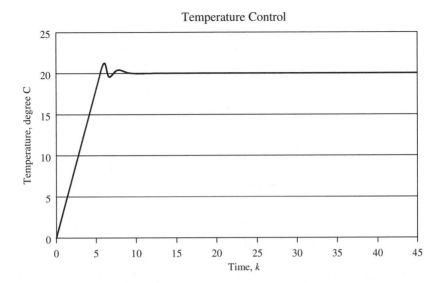

Figure 5.11 A simulated temperature tracking control result

In FLC design, another commonly used alternative for determining the weights of the control signal $u(t)$ is to give it a value of PL, PS, NL, or NS, for example, PL = 0.9, PS = 0.3, NL = −0.8 , NS = −0.2 . More specifically, a control rule base can be put in a tabular form as shown in Table 5.4.

Table 5.4 is sometimes called a "look-up table," convenient for the field engineers to use, which can be made more accurate by dividing both e and $\dot{e}$ into more subcases. Table 5.5 gives an example of a more subtle rule-base table for fuzzy logic controller. In these two tables,

$$\Delta u((k+1)T) = K \cdot \Delta u(kT) ,$$

where under the conditions on e and $\dot{e}$ to be PL, PS, etc., which do not take specific numerical values (see Table 5.6 in Example 5.1 below), the corresponding K takes the pre-assigned numerical values PL, PS, NS, NL (e.g., PL = 0.9, PS = 0.3, NL = −0.8 , NS = −0.2) given inside the tale, in which NM and ZO mean "negative medium" (e.g., NM = −0.5) and "zero," respectively, and the other abbreviations are similarly understood.

Table 5.4 A Rule Base in the Tabular Form for K

e \ $\dot{e}$	< 0	> 0
PL	PL	NL
PS	PS	NS
NS	NS	PS
NL	NL	PL

Table 5.5 A Rule Base Table for K

e \ $\dot{e}$	NL	NM	NS	ZO	PS	PM	PL
NL	NL	NL	NL	NL	NM	ZO	PS
NM	NL	NL	NL	NM	ZO	PS	PM
NS	NL	NL	NM	ZO	PS	PM	PL
ZO	NL	NM	ZO	PS	PM	PL	PL
PS	NM	ZO	PS	PM	PL	PL	PL
PM	ZO	PS	PM	PL	PL	PL	PL
PL	PS	PM	PL	PL	PL	PL	PL

(2.4) Establishing a fuzzy logic inference engine

In order to complete the fuzzy logic inference embedded in the control rule base, the general fuzzy IF-THEN rule has to be applied to each rule in the rule base.

Return to the Executive Rule Base discussed above, and take the first rule as an example:

R^1:   IF $e(kT) = $ PL AND $\dot{e}(kT) < 0$

THEN $u((k+1)T) = \mu_{PL}(e(kT)) \cdot u(kT)$.

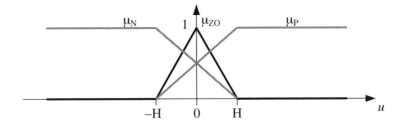

Figure 5.12 Typical membership functions for u

In this rule, $e(kT) = $ PL has a membership function $\mu_{PL}(e(kT))$ shown in Figure 5.10 (a), $\dot{e}(kT) < 0$ is nonfuzzy and so has membership values $\mu_N(\dot{e}(kT)) = 1$, and $\Delta u((k+1)T)$ has three membership functions as shown in Figure 5.7.

Since each rule will produce one control output, to obtain the final controller's output, it is common to use the output membership functions shown in Figure 5.12, along with the general weighted average formula, as shown in formula (5.4) below.

(3) The Defuzzification Module

The defuzzification module is in a sense the reverse of the fuzzification module: it converts all the fuzzy terms created by the rule base of the controller to crisp terms (numerical values) and then sends them to the physical system (plant, process), so as to execute the control of the system.

The defuzzification module performs the following functions:

(3.1) Establish output membership functions, typically they are as shown in Figure 5.12.

(3.2) Output a crisp, overall control signal, u, by combining all possible control outputs from the rule base into a recursive control action with a weighted average formula:

$$u(kT) = \frac{\sum\limits_{i=1}^{N} \mu_i \, u_i(kT)}{\sum\limits_{i=1}^{N} \mu_i} \tag{5.4}$$

where u_i is the output from rule R^i, $i = 1,..., N$, and μ_i, $i = 1,..., N$, indicate the output membership values given by the functions μ_N, μ_{ZO}, μ_P shown in Figure 5.11, which are simple, yet typical choices for the membership functions of u, with P, N, and ZO indicating positive, negative, and zero, respectively, and H is a real number selected by the designer or determined by tuning.

In the above weighted average formula, if, for instance, the output $u_1(kT) = 0.5H$ from rule R^1 then it has three output membership values: $\mu_N = 1/3$, $\mu_{ZO} = 1/2$, $\mu_P = 2/3$, therefore,

$$u(kT) = \frac{\mu_N \cdot u_1(kT) + \mu_{ZO} \cdot u_1(kT) + \mu_P \cdot u_1(kT) + \cdots}{\mu_N + \mu_{ZO} + \mu_P + \cdots}$$

$$= \frac{(1/3 + 1/2 + 2/3) \cdot u_1(kT) +}{1/3 + 1/2 + 2/3 + \cdots}$$

This is the same in using the weighted average formula as before.

Just like the fuzzification module, this step of the defuzzification module transforms the overall control output, u, obtained in the previous step, to the corresponding physical values (position, voltage, degree, etc.) that the given system (plant, process) can accept. This converts the fuzzy logic controller's numerical output to a physical means that can actually drive the given system to produce the expected outputs.

The overall fuzzy logic controller that combines the "fuzzification - rule base - defuzzification" modules has been shown in Figure 5.7, which is the controller block in the closed-loop set-point control system of Figure 5.6.

As another simple example, consider the temperature control problem illustrated by Figure 5.9, and suppose that the plant of the temperature control process as shown in Figure 5.6 is given by

$$y((k+1)T) = y(kT) + a \cdot u(kT)$$

where a is a constant, with initial condition $y(0) = 0$. This simple linear plant model is used to generate data for simulation.

Using the "Executive Rule Base" established above, with a slight change of the control actions in the following form

$$u((k+1)T) = u(kT) + \text{sign}(u(kT)) \cdot C$$

where $C > 0$ is a constant control input, the simulation results obtained in Figure 5.13 (a)-(c), with $T = 1$ for simplicity in simulation, clearly show that the temperature control process works quite well without knowing the mathematical model of the process.

This example also shows that, just like in the conventional controllers design, a controller for a given plant is usually not unique, and an experienced designer usually can come out with a simple controller with good performance.

Finally, it is very important to recall that the design of the fuzzy logic controller discussed in this section does not require any information about the system (plant, process): it doesn't matter if the system is linear or nonlinear (nor the order of linearity, structure of nonlinearity, etc.) as long as its output, y, can be measured by a sensor and used for the control. In other words, the design of the fuzzy logic controller is independent of the mathematical model of the system under control. Hence, this is referred to as the model-free approach to FLC design.

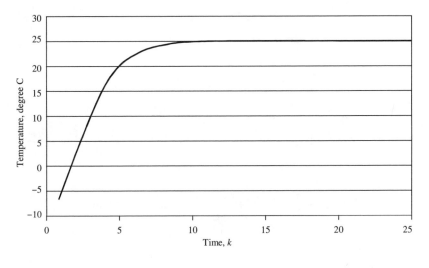

(a) $a = 13.5, \ H = 10, \ C = 2$

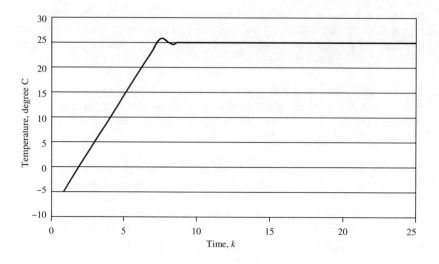

(b) $a = 13.5$, $H = 5$, $C = 2$

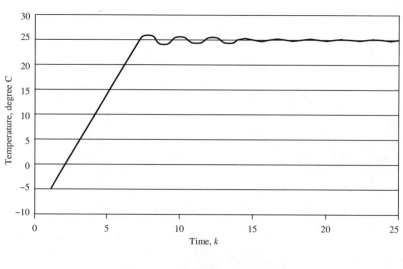

(c) $a = 13.5$, $H = 3$, $C = 2$

Figure 5.13 Simulation results on temperature control

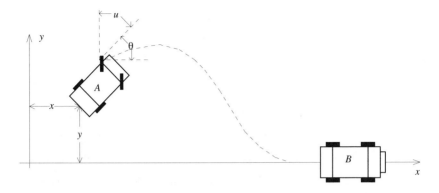

Figure 5.14 A truck-driving control example

Comparing it with the conventional controllers design, its advantages are obvious. The price that one has to pay for this success is the complexity in the design, i.e., the "fuzzification - rule base - defuzzification" steps. But this is what it is supposed to be: under a worse condition (the system model is unknown, or only vaguely given), if one wants to design a controller to achieve similar set-point tracking performance, then one has to do more analysis and calculation in the design.

C. Examples of Model-Free Fuzzy Controller Design

Two examples are given in this section to illustrate the model-free fuzzy control approach and to demonstrate its effectiveness in controlling a complex mechanical system without assuming its mathematical model.

Example 5.1

Consider a truck-parking control problem, as shown in Figure 5.14.

The objective is to design a fuzzy logic controller, without assuming a mathematical model of the truck, to park the truck anywhere on the x-axis. Suppose that the truck can move forward at a constant speed of $v = 0.5$ *m/s* and it is assumed that the truck is equipped with sensors that can measure location (x,y) and orientation (angle) θ at all times. The fuzzy logic controller is to provide an input, u, to rotate the steering wheels and, consequently, to maneuver the truck.

In the simulation, the input variables are the truck angle θ and the vertical position coordinate, y, while the output variable is the steering angle (signal), u. The variable ranges are pre-assigned as

$$-100 \leq y \leq 100, \quad -180° \leq \theta \leq 180°, \quad -30° \leq u \leq 30°.$$

Here, clockwise rotations are considered positive in θ, and, therefore, counterclockwise are negative. The linguistic terms used in this design are given as follows:

Angle θ	y-position	Steering angle u
AB: Above	AO: Above much	NB: Negative big
AC: Above center	AR: Above	NM: Negative medium
CE: Center	AH: Above horizontal	NS: Negative small
BC: Below center	HZ: Horizontal	ZE: Zero
BE: Below	BH: Below horizontal	PS: Positive small
BR: Below	PM: Positive medium	
BO: Below much	PB: Positive big	

The first step is to choose membership functions. They are as shown in Figures 5.15-17. Note that narrow membership functions are used to permit fine control near the designated parking spot, while wide membership functions are used to perform fast controls when the truck is far away from the parking place.

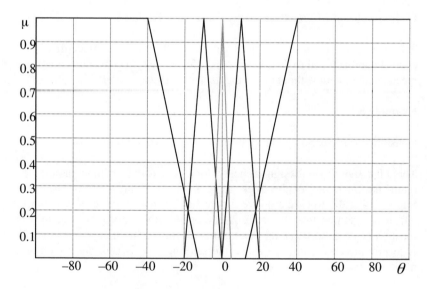

Figure 5.15 Fuzzy membership functions for the angle θ

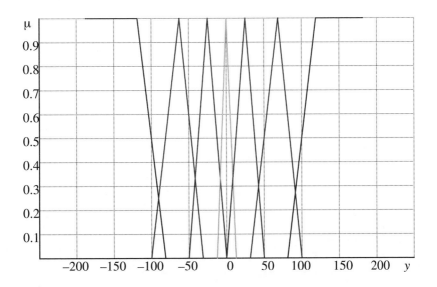

Figure 5.16 Fuzzy membership functions for the y position

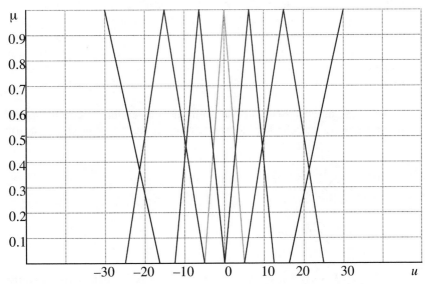

Figure 5.17 Fuzzy membership functions for the control signal u

Table 5.6 Fuzzy Controller Rule Base

	BE	BC	CE	AC	AB
BO	*PB*	*PB*	*PM*	*PM*	*PS*
BR	*PB*	*PB*	*PM*	*PS*	*NS*
BH	*PB*	*PM*	*PS*	*NS*	*NM*
HZ	*PM*	*PM*	*ZE*	*NM*	*NM*
AH	*PM*	*PS*	*NS*	*NM*	*NB*
AR	*PS*	*NS*	*NM*	*NB*	*NB*
AO	*NS*	*NM*	*NM*	*NB*	*NB*

The rule base used in simulation is summarized in Table 5.6, where PB = 0.9, PM = 0.5, PS = 0.3, NS = −0.3 , NM = −0.5 , NB = −0.3 , NB = − 0.9. Each rule has the form IF y is Y AND θ is Θ THEN u is U, as usual. A glance of these rules reveals the symmetry, an intrinsic property of this controller, which is reasonable for this parking application since the truck can move in any direction. This is not always the case, however, as can be seen from the next example where symmetry of control rules is not reasonable.

Finally, the output action, with the given input conditions is constructed. The following weighted average defuzzification formula was used for the final control output in simulation:

$$u = \frac{\sum_{i=1}^{8} \mu(u_i)u_i}{\sum_{i=1}^{8} \mu(u_i)} .$$

A total of 12 computer simulations were performed. Their initial conditions are summarized in Table 5.7, and the corresponding parking performances are shown in Figure 5.19.

Table 5.7 Initial Conditions of the 12 Simulated Cases

Case	1	2	3	4	5	6	7	8	9	10	11	12
θ	0	0	0	90	90	90	180	180	180	–90	–90	–90
y	30	20	–20	30	10	–20	30	–20	–10	–10	–20	20

The following is a simplified mathematical model for the motion of the truck. This model was used to create data for simulation but was not used in the design of the controller. This model directly follows from the geometry of the truck (see Figure 5.18):

$$\theta(k+1) = \theta(k) + v\,T\tan(\,u(k)\,)\,/\,L,$$

$$x(k+1) = x(k) + v\,T\cos(\,\theta(k)\,), \qquad\qquad (5.11)$$

$$y(k+1) = y(k) + v\,T\sin(\,\theta(k)\,),$$

where

$\theta(k)$ — angle of the truck at time k,

$x(k)$ — horizontal position of the truck rear-end at time k,

$y(k)$ — vertical position of the truck rear-end at time k,

$u(k)$ — steering angle as control input to the truck at time k,

L — length of the truck (L=2.5m),

T — sampling time of the discrete model (T=0.1s),

v — constant speed of the truck (v=0.5m/s).

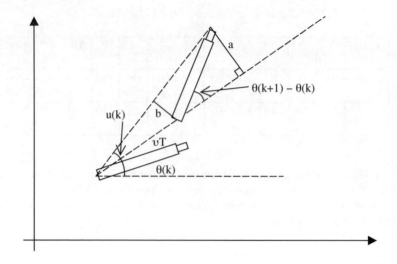

Figure 5.18 Geometry of the truck model

It can be seen from the figure that

$$\begin{cases} \sin\big[\theta(k+1)-\theta(k)\big] = a/L \\ \quad \tan[u(k)] = b/(vT) \end{cases}$$

For small movement, with a small L, one has

$$b \approx a \quad \text{and} \quad \sin[\theta(k+1)-\theta(k)] \approx \theta(k+1)-\theta(k),$$

yielding the truck model (5.11).

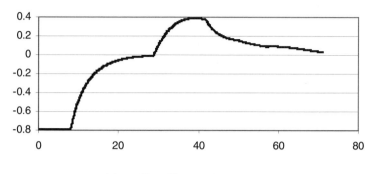

(a) profile of input $u(t)$ vs. time t

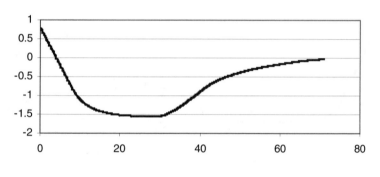

(b) profile of output $\theta(t)$ vs. time t

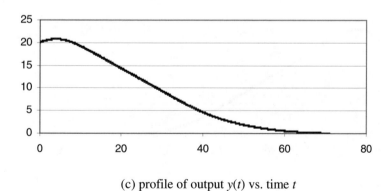

(c) profile of output $y(t)$ vs. time t

Figure 5.19 Model-free parking control simulation results

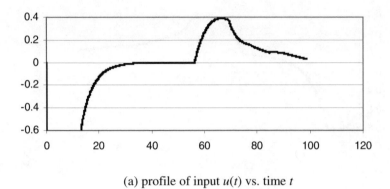

(a) profile of input $u(t)$ vs. time t

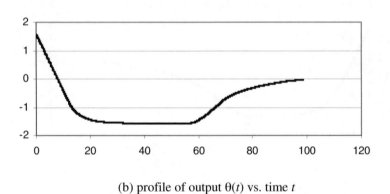

(b) profile of output $\theta(t)$ vs. time t

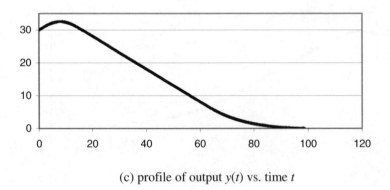

(c) profile of output $y(t)$ vs. time t

Figure 5.19 Model-free parking control simulation results (continued)

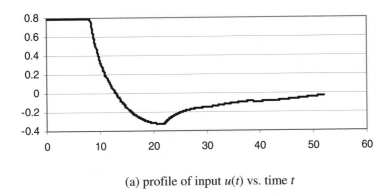

(a) profile of input $u(t)$ vs. time t

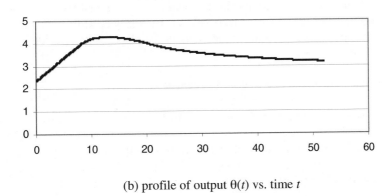

(b) profile of output $\theta(t)$ vs. time t

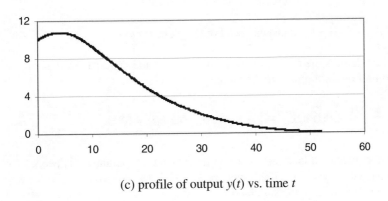

(c) profile of output $y(t)$ vs. time t

Figure 5.19 Model-free parking control simulation results (continued)

Example 5.2

A landing control problem of a model plane is discussed from a model-free approach. This airplane landing problem is visualized by Figure 5.20, which differs from the truck-parking control problem of Example 5.1 in that the airplane is not allowed to "overshoot" the set-point (i.e., cannot hit the ground) in order to avoid crashing. Therefore, the control rule base will not have symmetry in this example.

The objective is to design a fuzzy logic controller to land the plane safely, anywhere on the ground (the x-axis). Assume that no mathematical model for the plane is available for the design, but the plane is equipped with sensors that can measure the height $h(t)$ and the angle θ of the plane. Suppose that the model plane is moving forward at a constant speed of $v=5$ m/s, and the controller is used to steer the angle of the motion. Let the initial conditions be $h(0) = 1000m$ and $\theta(0) = 0$ $deg.$ (horizontally). The fuzzy controller is used to provide an input, u, which controls the angle of the plane, so as to guide it to land on the ground safely.

The parameters involved in the control system are as follows:

$\theta(t)$ — angle of the plane at time t,

$h(t)$ — vertical position of the plane at time t,

$u(t)$ — control input to the plane at time t,

T — sampling time for discretization ($1s$),

v — constant speed of the plane ($5m/s$).

One first defines the ranges of the plane position and angle, as well as that of the control input, throughout the landing process:

$$-\frac{\pi}{4} \le \theta(t) \le 0, \quad -\frac{\pi}{4} \le u(t) \le 0, \quad 0 \le h(t) \le 1000.$$

Since the plane can land anywhere on the ground, no parameter is needed for the horizontal axis. The final state is $h = 0 \pm 1m$ and $\theta = 0 \pm 0.1$ $deg.$ (horizontally), to ease the simulation.

The following values are chosen for fuzzification:

Height

h	Z	PS	PM	PL
Value	0	300	600	1000

Angle

θ	NL	NM	NS	Z
Value	$-\pi/4$	$-\pi/6$	$-\pi/8$	0

The membership functions for the height h and the angle θ are chosen as shown in Figures 5.21 and 5.22, respectively.

The fuzzy rule base used in the simulation is shown in Table 5.8, where V means "very" with PMS = 0.4, PS = 0.3, PVS = 0.1, NVS = –0.1, NS = –0.3, NMS = –0.4. It is also in the form of

   "IF h is H AND θ is Θ THEN u is U"

Table 5.8 Fuzzy Control Rule Base

h \ θ	NL	NM	NS	Z
Z	PMS	PS	PVS	Z
PS	PS	PVS	Z	NVS
PM	PVS	Z	NVS	NS
PL	Z	NVS	NS	NMS

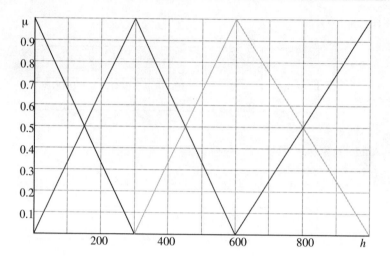

Figure 5.21 Fuzzy membership function for the height (h)

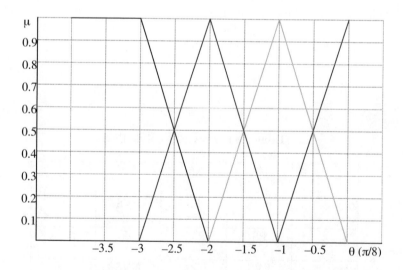

Figure 5.22 Fuzzy membership function for the angle (θ)

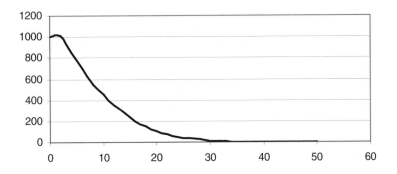

Figure 5.23 Airplane-landing control simulation result

The basic idea in constructing this simple control rule base is as follows: First, if the plane is far away from the ground, the degree of the plane angle (downward) is increased. Second, as the plane approaches the ground, the desired plane angle will gradually turn from downward to horizontal, so that it can land smoothly and safely.

The defuzzification formula used is, again, the weighted average of all control actions, where the weights are the corresponding membership values:

$$u = \frac{\displaystyle\sum_{i=1}^{n} \mu(u_i) u_i}{\displaystyle\sum_{i=1}^{n} \mu(u_i)}.$$

For the purpose of computer simulation, the following simplified mathematical model of the plane was used to create data for control (but it was not used in the controller design):

$$\theta(k+1) = \theta(k) + vT \tan(u(k))$$

$$y(k+1) = y(k) + vT \sin(u(k)). \tag{5.12}$$

A simulation result is shown in Figure 5.23, which demonstrates the safe landing process of the plane under the designed fuzzy logic controller.

Problems

P5.1   A conventional machine tool working in an automatic cycle is under the control of a programmable logic controller (PLC). This machine tool is loaded and unloaded by an industrial robot arm that operates according to a 2-second load cycle and then a 2-second unload cycle. There are several other activities to be performed at scheduled sequential times, some of them overlapping, as summarized in Figure 5.24.

Fill out the drum-timer array displayed in Table 5.9 for the machine tool.

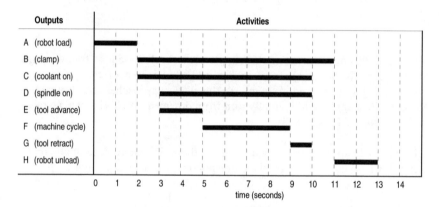

Figure 5.24 Activities of the machine tool

Table 5.9 PLC Drum-Timer Array for the Machine Tool

Steps	Counts (1 count/sec)	Outputs							
		A	B	C	D	E	F	G	H
1	2								
2	1								
3	2								
4	4								
5	1								
6	1								
7	2								

P5.2   Consider a car-parking control application using fuzzy logic (see Figure 5.25). The objective is to design a fuzzy controller, without assuming a mathematical model of the car, to park the car anywhere on the y-axis. Suppose that the car can move forward at a constant speed and that the car is equipped with sensors that can measure location (x,y) and orientation (angle) θ at all times. The fuzzy logic controller is to provide an input, u, to rotate the steering wheels and, consequently, to maneuver the car. Write a simple yet complete design with reasonable membership functions and a working rule base for the fuzzy controller.

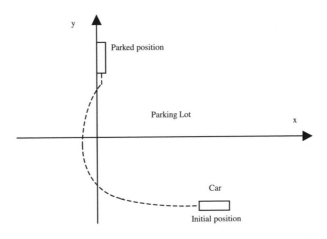

Figure 5.25 Car parking control

CHAPTER 6

Fuzzy PID Control Systems

In the last section, one has seen that fuzzy logic controllers can be designed to work even without any knowledge about the mathematical model (differential equation, transfer function, etc.) of the system (plant, process) for set-point tracking problems if the system output can be measured and used on-line. The main idea was to use the error signal (the difference between the reference signal and the system output) to drive the fuzzy logic controller, so as to create a new control action each time. These control actions should be able to alter the system outputs in such a way that the error signal is being continuously reduced until a satisfactory set-point tracking performance is achieved.

This section provides the studies of another type of fuzzy controllers: the fuzzy PID controllers.

Conventional (or classical) proportional-integral-derivative (PID) controllers are the most well-known and most widely used controllers in modern industries: statistics have shown that more than 90% controllers used in industries today are PID or PID-type of controllers, and the rest are mainly programmable logic controllers (PLCs), which were discussed at the beginning of the last chapter.

Generally speaking, PID controllers have the merits of being simple, reliable, and effective: they consume lower cost but are very easy to operate. Besides, for lower-order linear systems (plants, processes), PID controllers have remarkable set-point tracking performance with guaranteed stability. Therefore, PID controllers are very popular in real-world applications.

I. CONVENTIONAL PID CONTROLLERS

First, a brief review on several basic types of conventional PID controllers is given, on their configurations and design methods.

Individually, the conventional proportional (P), integral (I), and derivative (D) controllers for controlling a given system (plant, process) have the structures shown in Figure 6.1 (a), (b), and (c), respectively, where $r = r(t)$ is the reference input (set-point), $y = y(t)$ is the controlled system's output, $e = e(t) := r(t) - y(t)$ is the set-point tracking error, and $u = u(t)$ is the control action (output of the controller) which is used as the input to the plant.

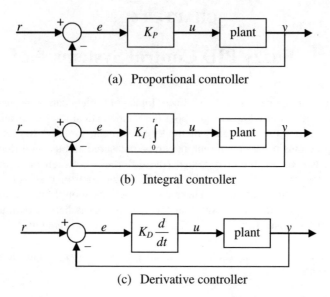

(a) Proportional controller

(b) Integral controller

(c) Derivative controller

Figure 6.1 Proportional-integral-derivative controllers

In the time domain, they have the following representations:

(i)   The P-controller $u(t) = K_P\, e(t)$;

(ii) The I-controller $u(t) = K_I \int_0^t e(\tau)\, d\tau$;

(iii) The D-controller $u(t) = K_D\, \dfrac{d}{dt}\, e(t)$.

Here, the three control gains, K_P, K_I, and K_D, are constants to be determined in the design according to the set-point tracking performance and stability consideration.

To increase the control capabilities of these simple individual controllers, they are usually used in combinations, which will be further discussed below.

The following are some basic, typical combinations of P, I, and D controllers (see Figure 6.2 for their configurations):

(iv) PI-controller $u(t) = K_P\, e(t) + K_I \int_0^t e(\tau)\, d\tau$;

(v)   PD-controller $\qquad u(t) = K_P\,e(t) + K_D\,\dfrac{d}{dt}\,e(t);$

(vi)   PID-controller $\qquad u(t) = K_P\,e(t) + K_I \displaystyle\int_0^t e(\tau)\,d\tau + K_D\,\dfrac{d}{dt}\,e(t);$

(vii)   PI+D-controller $\qquad u(t) = K_P\,e(t) + K_I \displaystyle\int_0^t e(\tau)\,d\tau - K_D\,\dfrac{d}{dt}\,y(t).$

In the frequency domain, one has the following corresponding representations for the individual P, I, and D controllers:

(i)   The P-controller $\qquad U(s) = K_P\,E(s);$

(ii)   The I-controller $\qquad U(s) = \dfrac{K_I}{s}\,E(s);$

(iii)   The D-controller $\qquad U(s) = K_D\,s\,E(s).$

Here, as usual, capital letters are used for the Laplace transform $L\{\cdot\}$ of a continuous-time signal, or the z-transform $Z\{\cdot\}$ of a discrete-time signal. Thus, with zero initial conditions,

$$U(s) = L\{\,u(t)\,\} \quad \text{and} \quad E(s) = L\{\,e(t)\,\}.$$

The following are some basic, typical combinations of P, I, and D controllers in the frequency domain:

(iv)   PI-controller $\qquad U(s) = K_P\,E(s) + \dfrac{K_I}{s}\,E(s);$

(v)   PD-controller $\qquad U(s) = K_P\,E(s) + K_D\,s\,E(s);$

(vi)   PID-controller $\qquad U(s) = K_P\,E(s) + \dfrac{K_I}{s}\,E(s) + K_D\,s\,E(s);$

(vii)   PI+D-controller $\qquad U(s) = K_P\,E(s) + \dfrac{K_I}{s}\,E(s) - K_D\,s\,Y(s).$

In the above, one has

$$Y(s) = L\{\,y(t)\,\} = L\{\,r(t) - e(t)\,\} = R(s) - E(s).$$

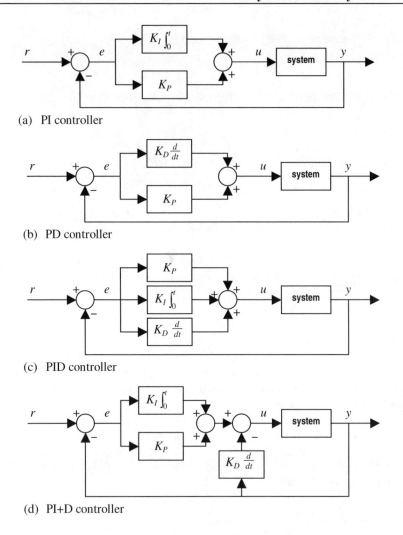

(a) PI controller

(b) PD controller

(c) PID controller

(d) PI+D controller

Figure 6.2 Some typical combinations of P, I, D controllers

Note that the PID controller shown in Figure 6.2 (c) is not a good combination of the three controllers in practice, since I and D controllers have reverse actions (so they will not be combined as an ID controller in general). The combination shown in Figure 6.2 (c) is good only for illustration since it has a neat formulation and, hence, is called a "textbook PID controller." A practical combination of the three controllers is the PI+D controller shown in Figure 6.2 (d), where the system output signal y(t) is usually smoother than the error signal (through the I-controller and the plant), or a PD+I controller in a similar configuration. More importantly, this structure of closed-loop control systems

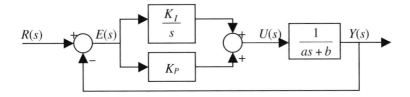

Figure 6.3 A PI control system

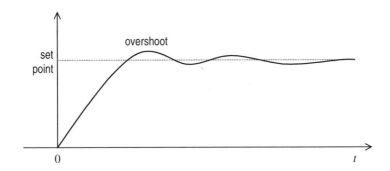

Figure 6.4 A typical tracking performance of the PI controller

has been validated to be efficient by many practical examples and case studies.

To show how a PID-type of controller works and how to design such a controller to perform set-point tracking with guaranteed stability, two simple examples are first discussed below.

Example 6.1

Consider the PI-control system shown in the frequency domain by Figure 6.3, where the given linear system has a first-order transfer function, $1/(as + b)$, with known constants $a > 0$ and b.

The reference r is a given constant (set-point). The design of the PI controller is to determine the two constant control gains, K_P and K_I, such that the system output $y(t)$ can track the reference:

$$y(t) \rightarrow r \text{ as } \qquad t \rightarrow \infty,$$

while the entire feedback control system is stable even if the given transfer function $1/(as + b)$ is unstable.

First, observe from Figure 6.3 that

$$U(s) = K_P\, E(s) + \frac{K_I}{s}\, E(s),$$

$$E(s) = R(s) - Y(s),$$

$$Y(s) = \frac{1}{as+b}\, U(s).$$

By combining these relations together, one can obtain the following overall closed-loop system input-output relation:

$$Y(s) = H(s)\, R(s) = \frac{K_P s + K_I}{as^2 + (b + K_P)s + K_I}\, R(s), \qquad (6.1)$$

where $H(s)$ is the transfer function of the overall closed-loop feedback control system. It is easy to see that this transfer function has two poles:

$$s_{1,2} = \frac{-(b + K_P) \pm \sqrt{(b + K_P)^2 - 4aK_I}}{2a}. \qquad (6.2)$$

Thus, in this design, if one chooses (note: $a > 0$)

$$K_P > -b, \qquad (6.3)$$

then one can guarantee that these two poles have negative real parts, so that the overall closed-loop controlled system is stable.

For set-point tracking, the goal is

$$\lim_{t \to \infty} e(t) = 0. \qquad (6.4)$$

It follows from the standard Terminal-Value Theorem of Laplace transforms that

$$\lim_{t \to \infty} e(t) = \lim_{|s| \to 0} s\, E(s)$$

$$= \lim_{|s| \to 0} s\, [\, R(s) - Y(s)\,]$$

$$= \lim_{|s| \to 0} s\, [\, 1 - H(s)\,]\, R(s)$$

$$= \lim_{|s| \to 0} s \cdot \frac{as^2 + bs}{as^2 + (b + K_\mathrm{P})s + K_\mathrm{I}} \cdot \frac{r}{s}$$

$$= 0.$$

It may appear that the set-point tracking task can always be done no matter how one chooses the control gain K_I, provided that the other control gain, K_P, satisfies the condition (6.3), and this is true for any given constants a and b in the given plant. This observation is correct for this example. However, in so doing it often happens that the system output $y(t)$ tracks the set-point r with higher-frequency oscillations caused by the pure imaginary parts of the two poles given in (6.2). Hence, to eliminate such undesirable oscillations so as to obtain better tracking performance, one can select the control gain K_I to zero out the pure imaginary parts of the poles s_1 and s_2. Namely, one can force $(b + K_\mathrm{P})^2 - 4\,a\,K_\mathrm{I} = 0$, which yields

$$K_\mathrm{I} = \frac{(b + K_\mathrm{P})^2}{4a},$$

where K_P has been determined by (6.3). Thus, the design of the PI controller is completed, for the given plant $1/(as + b)$, which can be originally unstable (i.e., $b < 0$). One thus obtains an overall stable feedback control system whose output can track the set-point without oscillations, at least in theory. Its output $y(t)$ generally has the shape as shown in Figure 6.4.

As seen, in this example a PI controller so designed can completely eliminate both the steady-state tracking error and the transient oscillations, but may not be able to reduce the maximum overshoot in the output since there is no more degree of freedom to do so. This observation on PI controllers is generally true.

On the contrary, a PD controller can generally improve the tracking performance by reducing the maximum overshoot of the output, but may not be able to eliminate the steady-state tracking error, as can be seen from the next example.

Example 6.2

Consider the PD-control system shown in the frequency domain by Figure 6.5, where the given linear plant has a second-order transfer function, $1/(as^2 + bs + c)$, with known constants $a > 0$, b, and c. The design of the PD controller is to determine the two constant control gains, K_P and K_D, such that the system

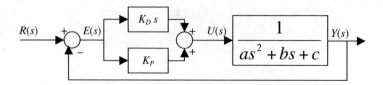

Figure 6.5 A PD control system

output can track the set-point while the entire feedback control system is stable even if the given transfer function $1/(as^2 + bs + c)$ is unstable.

It follows from Figure 6.5 that

$$U(s) = K_P E(s) + K_D s E(s),$$

$$E(s) = R(s) - Y(s),$$

$$Y(s) = \frac{1}{as^2 + bs + c} U(s),$$

so that

$$Y(s) = H(s) R(s) = \frac{K_P + K_D s}{as^2 + (b + K_D)s + (c + K_P)} R(s).$$

The transfer function $H(s)$ of the overall feedback control system has two poles:

$$s_{1,2} = \frac{-(b + K_D) \pm \sqrt{(b + K_D)^2 - 4a(c + K_P)}}{2a}, \tag{6.5}$$

and so the selection (note $a > 0$)

$$K_D > -b \tag{6.6}$$

and

$$K_P = \frac{(b + K_D)^2}{4a} - c \tag{6.7}$$

can guarantee the controlled system be stable and have no oscillations on the output trajectory during the set-point tracking process, at least in theory. However, the asymptotic tracking error for this PD controlled system is

$$\lim_{t \to \infty} e(t) = \lim_{|s| \to 0} s\, E(s)$$

$$= \lim_{|s| \to 0} s[I - H(s)]R(s)$$

$$= \lim_{|s| \to 0} s \cdot \frac{as^2 + bs + c}{as^2 + (b + K_{\mathrm{D}})s + (c + K_{\mathrm{P}})} \cdot \frac{r}{s}$$

$$= \frac{cr}{c + K_{\mathrm{P}}},$$

which is not zero if $c \neq 0$ and $r \neq 0$, meaning that the PD controller generally cannot eliminate the steady-state error in set-point tracking. Of course, this tracking error can be reduced by using a large value (high gain) of K_{P} within the limit of physical implementation.

The PD controller, as compared to the PI controller, has its advantages: it can produce smaller maximum overshoot and is more sensitive (easier to tune) in general, as widely experienced.

II. FUZZY PID CONTROLLERS (TYPE-1)

To implement a fuzzy PI or PD controller on a computer, one first needs the digital version of the analog one discussed above.

A. Discretization of PID Controllers

To digitize an analog controller, the following bilinear transform for discretization is usually used:

$$s = \frac{2}{T} \frac{z - 1}{z + 1} \tag{6.8}$$

where $T > 0$ is the sampling time.

For the continuous-time PI controller, one obtains (see Figure 6.3):

$$\frac{K_I}{s} \rightarrow \frac{K_I T}{2} \frac{z+1}{z-1} = \frac{K_I T}{2} \frac{2-(1-z^{-1})}{1-z^{-1}} = -\frac{K_I T}{2} + \frac{K_I T}{1-z^{-1}},$$

$$U(s) = \left(K_P + \frac{K_I}{s} \right) E(s) \rightarrow \tilde{U}(z) = \left(K_P - \frac{K_I T}{2} + \frac{K_I T}{1-z^{-1}} \right) \tilde{E}(z).$$

Let

$$\tilde{K}_P = K_P - \frac{K_I T}{2} \qquad \text{and} \qquad \tilde{K}_I = K_I T. \tag{6.9}$$

Then,

$$(1 - z^{-1})\tilde{U}(z) = \tilde{K}_P (1 - z^{-1}) \tilde{E}(z) + \tilde{K}_I \tilde{E}(z).$$

It then follows from the inverse z-transform that

$$u(nT) - u(nT{-}T) = \tilde{K}_P [e(nT) - e(nT{-}T)] + \tilde{K}_I e(nT),$$

so that

$$\frac{u(nT) - u(nT-T)}{T} = \tilde{K}_P \frac{e(nT) - e(nT-T)}{T} + \frac{\tilde{K}_I}{T} e(nT).$$

Let, furthermore,

$$\Delta u(nT) = \frac{u(nT) - u(nT-T)}{T} \tag{6.10}$$

be the incremental control and

$$v(nT) = \frac{e(nT) - e(nT-T)}{T}. \tag{6.11}$$

One thus has the following iterative formula:

$$u(nT) = u(nT{-}T) + T\Delta u(nT), \tag{6.12}$$

with

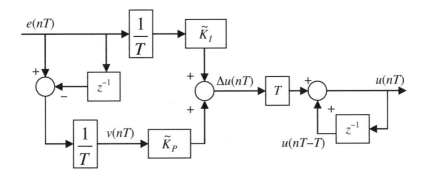

Figure 6.6 The digital PI controller

$$\Delta u(nT) = \tilde{K}_{\mathrm{P}}\, v(nT) + \frac{\tilde{K}_{\mathrm{I}}}{T}\, e(nT), \tag{6.13}$$

which can be implemented as shown in Figure 6.6.

Similarly, using the bilinear transform (6.8) in the continuous-time PD controller (see Figure 6.5), one obtains

$$s\,K_{\mathrm{D}} \rightarrow \frac{2}{T}\frac{z-1}{z+1}K_{\mathrm{D}} = \frac{2}{T}\frac{1-z^{-1}}{1+z^{-1}}K_{\mathrm{D}},$$

and

$$U(s) = (K_{\mathrm{P}} + sK_{\mathrm{D}})\,E(s) \rightarrow \tilde{U}(z) = \left(K_{P} + \frac{2}{T}\frac{1-z^{-1}}{1+z^{-1}}K_{D} \right)\tilde{E}(z).$$

Let

$$\tilde{K}_{\mathrm{P}} = K_{\mathrm{P}} \qquad \text{and} \qquad \tilde{K}_{\mathrm{D}} = \frac{2}{T}K_{\mathrm{D}}. \tag{6.14}$$

Then,

$$(1 + z^{-1})\tilde{U}(z) = (1 + z^{-1})\,\tilde{K}_{\mathrm{P}}\,\tilde{E}(z) + \tilde{K}_{\mathrm{D}}\,(1 - z^{-1})\,\tilde{E}(z),$$

so that the inverse z-transform gives

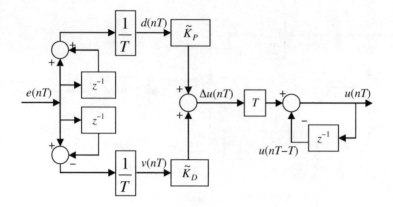

Figure 6.7 The digital PD controller

$$\Delta u(nT) = \tilde{K}_P \, d(nT) + \tilde{K}_D \, v(nT), \qquad\qquad (6.15)$$

where

$$d(nT) = \frac{e(nT) + e(nT - T)}{T},$$

$$v(nT) = \frac{e(nT) - e(nT - T)}{T}.$$

This digital PD controller (6.15) can be implemented as shown in Figure 6.7.

Note that formula (6.15) can be rewritten as

$$u(nT) = - u(nT{-}T) + \tilde{K}_P \, [e(nT) + e(nT{-}T)] + \tilde{K}_D \, [e(nT) - e(nT{-}T)],$$

and, consequently,

$$u(nT{-}T) = - u(nT{-}2T) + \tilde{K}_P \, [e(nT{-}T) + e(nT{-}2T)]$$

$$+ \tilde{K}_D \, [e(nT{-}T) - e(nT{-}2T)].$$

Thus, by substituting $u(nT{-}T)$ into $u(nT)$, one obtains

$$u(nT) \;=\; u(nT\!-\!2T) + \tilde{K}_P\,[e(nT) - e(nT\!-\!2T)]$$

$$+\; \tilde{K}_D\,[e(nT) - 2e(nT\!-\!T) + e(nT\!-\!2T)]$$

$$=\; u(nT\!-\!2T) + \tilde{K}_P\,T\,\frac{e(nT) - e(nT-2T)}{T}$$

$$+\tilde{K}_D\,T^2\,\frac{e(nT) - 2e(nT-T) + e(nT-2T)}{T^2}\,,$$

where the two finite divided differences are the discretization of $\dot{e}(t)$ and $\ddot{e}(t)$, respectively, of the continuous-time error signal $e(t)$, under the bilinear transform (6.8).

This formula shows clearly that the digital PD controller implicitly uses the discretization of both $\dot{e}(t)$ and $\ddot{e}(t)$, as is known in the conventional control theory. Since this formula is equivalent to, but more complicated than, formula (6.15), it will not be used in general.

B. Designing Type-1 Fuzzy PID Controllers

The fuzzy PI, PD, and even PID controllers can be designed by employing the digital PI controller (6.12)-(6.13) and the digital PD controller (6.15), with implementations as shown in Figures 6.6 and 6.7, respectively.

The type-2 fuzzy PID controllers use fuzzy logic rule base to tune the P, I, D control gains K_P, K_I, K_D on-line.

Since the P gain usually is used to control the rise time of the tracking error, a simple rule base may be established for it based on the following heuristics (principle):

- IF the tracking error is large THEN the P-gain is large

- IF the tracking error is small THEN the P-gain is small

As to the PI part of the controller, which can eliminate the steady-state tracking error, one may use the following principle for a rule base:

- F the steady-state tracking error is large THEN the I-gain is large

- F the steady-state tracking error is small THEN the I-gain is small

Table 6.1 Rule base for determining the gain K_P

e \ ė	NB	NS	ZE	PS	PB
NB	VB	VB	VB	VB	VB
NS	B	B	B	MB	VB
ZE	ZE	ZE	MS	S	S
PS	B	B	B	MB	VB
PB	VB	VB	VB	VB	VB

Similarly, since the PD part of the controller can eliminate the overshoot in the set-point tracking, its rule base may be established by following the following basic principle:

- IF the overshoot is large THEN the D-gain is large

- IF the overshoot is small THEN the D-gain is small

C. Two Examples

A simulated example is the following simple PID control system.

Example 6.3

Consider a plant with the transfer function $\dfrac{1}{s^2+10s+1}$. For simplicity here, the PID controller is directly discretized from the continuous-time version, and is obtained as

$$u(k) = K_P e(k) + K_I Te(k-1) + K_D[e(k) - e(k-1)]/T$$

where

$$Te(k-1) \approx \int_{(k-1)T}^{kT} e(\tau)d\tau \text{ and } [e(k) - e(k-1)]/T \approx \dot{e}(t),$$

Table 6.2 Rule base for determining the gain K_I

$\dot{e}$ e	NB	NS	ZE	PS	PB
NB	M	M	M	M	M
NS	S	S	S	S	S
ZE	MS	MS	ZE	MS	MS
PS	S	S	S	S	S
PB	M	M	M	M	M

Table 6.3 Rule base for determining the gain K_D

$\dot{e}$ e	NB	NS	ZE	PS	PB
NB	ZE	S	M	MB	VB
NS	S	B	MB	VB	VB
ZE	M	MB	MB	VB	VB
PS	B	VB	VB	VB	VB
PB	VB	VB	VB	VB	VB

in which $T > 0$ is the sampling time.

In this example, the rule base for determining control gain K_P is given in Table 6.1.

The rule base for determining control gain K_I is given in Table 6.2.

The rule base for determining control gain K_D is given in Table 6.3.

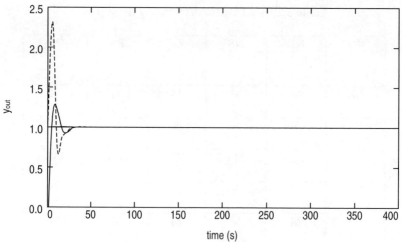

Figure 6.8 Control performance of the PID controller

With carefully chosen values of PB, VB, B, M, etc. in Tables 6.1 – 6.3, this controller works, although not very effectively.

A typical simulation result is shown in Figure 6.8, where the solid line is the system output and the dashed line is the error signal.

Example 6.4

Consider a typical fuzzy control system, as shown in Figure 6.9.

In this system, the plant is the chaotic Chua's circuit, given by (see Figure 6.10 (a)):

$$C_1 \dot{v}_{C_1} = R^{-1}(v_{C_2} - v_{C_1}) - f(v_{C_1})$$
$$C_2 \dot{v}_{C_2} = R^{-1}(v_{C1} - v_{C2}) + i_L \qquad (6.16)$$
$$L \dot{i}_L = -v_{C_2}$$

where L is a conductor, i_L is the current through L, C_1 and C_2 are two capacitors, v_{C_1} and v_{C_2} are the voltages across C_1 and C_2, respectively, R is a resistor, and R_N is a nonlinear resistor described by $f(\cdot)$ in (6.16) with expression

$$f(v_{C_1}) = m_0 v_{C_1} + \frac{1}{2}(m_1 - m_0) \left(|v_{C_1} + 1| - |v_{C_1} - 1| \right)$$

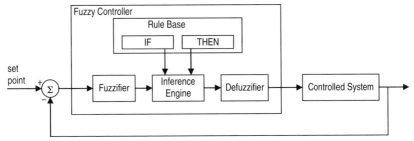

Figure 6.9 A typical fuzzy control system

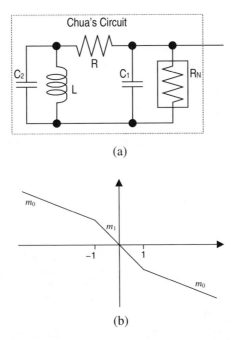

Figure 6.10 Chua's circuit and its nonlinear resistor

in which the two constants $m_0 < 0$ and $m_1 > 0$, as shown in Figure 6.10 (b).

This circuit can produce complex chaotic behaviors, as shown in Figure 6.11.

A fuzzy PID controller is to control the nonlinear resistor of the chaotic circuit (see Figure 6.12). In this figure, K_u, K_e and $K_{\Delta e}$ are constant gains to be adjusted in the control process, which can be regarded as PID control gains, respectively.

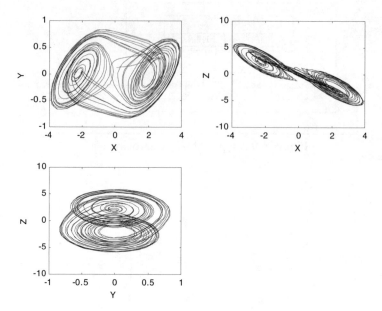

Figure 6.11 Complex behavior of Chua's circuit: A chaotic attractor

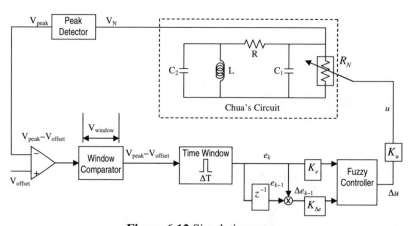

Figure 6.12 Simulation setup

In the controller, the rule base is given in Table 6.4, where the real values of PB, PS, etc. are given in Table 6.5 according to different levels of signal values (compare the two tables column-wise).

Table 6.4 Rule base for the controller

Δe \\ e	NB	NS	ZO	PS	PB
NB	NB	NS	NS	ZO	ZO
NS	NS	ZO	ZO	PS	/
ZO	NS	ZO	PS	PS	/
PS	ZO	ZO	PS	PB	/
PB	ZO	PS	PS	PB	/

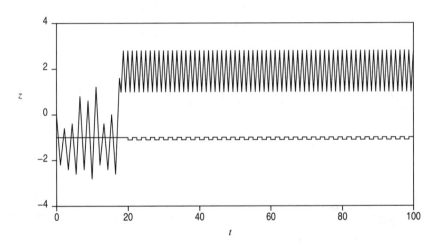

Figure 6.13 Chaotic signal is controlled to period-1 oscillation

A simulation result is shown in Figure 6.13, where the chaotic signal (circuit current i_L) is controlled to a period-1 oscillation. In this figure, the controller is switched on at time $t = 20$ and the lower curve is the control signal u generated by the controller.

Table 6.5 Values of PB, PM, PS, ZO, NS, NM, NB

System Variables	Signal Values								
		NB	NM	NS	ZO	PS	PM	PB	
E	−1	−0.75	−0.5	−0.25	0	0.25	0.5	0.75	1
Δe	−1	−0.75	−0.5	−0.25	0	0.25	0.5	0.75	1
Δu	−1	−0.75	−0.5	−0.25	0	0.25	0.5	0.75	1

Linguistic Variables	Membership Values for $K_u, K_e, K_{\Delta e}$								
PB	0	0	0	0	0	0	0	0.6	1
PM	0	0	0	0	0	0	0.6	1	0.6
PS	0	0	0	0	0	0.5	1	0.5	0
ZO	0	0	0	0.5	1	0.5	0	0	0
NS	0	0.5	1	0.5	0	0	0	0	0
NM	0.6	1	0.6	0	0	0	0	0	0
NB	1	0.6	0	0	0	0	0	0	0

III. FUZZY PID CONTROLLERS (TYPE-2)

Type-2 fuzzy PID controllers discussed in the last section, based on heuristic rule bases, are quite simple. However, they usually do not work very efficiently.

Another more advanced type of fuzzy PID controllers can be designed, which have precise analytic formulas along with rigorous stability analysis, and work very efficiently.

This type of fuzzy PID controllers, referred to as type-2 fuzzy PID controllers, is now introduced, including their design methods, performance evaluation, and stability analysis.

More specifically, the fuzzy PD controller design is first studied in detail, in which all the basic ideas, design principles, and step-by-step derivation and calculations are discussed. The fuzzy PI controllers design will be discussed briefly next, followed by the fuzzy PI+D controllers design. Having this background, many other types of fuzzy PID controllers can be designed by following similar procedures. The stability analysis of this type of fuzzy PID controllers will be investigated in the last section of the chapter.

Although it has been seen in the last section that it is possible to design type-1 fuzzy PID controllers by inserting some heuristic fuzzy logic IF-THEN rule bases into the control system to self-tune the PID control gains, that approach in general does not come up with new fuzzy PID controllers that capture the essential characteristics and nature of the conventional PID controllers. Besides, they generally do not have analytic formulas to use for control specification and stability analysis.

The type-2 fuzzy controllers, particularly the fuzzy PD, PI, and PI+D controllers to be introduced below, are natural extensions of their conventional versions, which preserve the linear structures of the PID controllers, with simple and conventional analytical formulas as the final results of the design. Thus, they can be directly used to replace the conventional PID controllers in any operating control system (plant, process). The main difference is that these fuzzy PID controllers are designed by employing fuzzy logic control principles and techniques, which have been studied in the last few chapters, to obtain new controllers that possess analytical formulas very similar to the conventional digital PID controllers. After the design is completed, all the fuzzy logic IF-THEN rules, membership functions, defuzzification formulas, etc. will no longer be needed in the applications of the controllers: what one can see is merely a conventional controller with a few simple formulas similar to the familiar conventional PID controllers. Thus, in operations a control engineer who does not have any knowledge about fuzzy logic and fuzzy control systems can use them just like the conventional ones, particularly for higher-order, time-delayed, and nonlinear systems, and for those systems that have only vague or incomplete mathematical models or contain significant uncertainties. For such systems, it turns out that fuzzy PID controllers work (much) better than their conventional counterparts.

The key reason, which also is the price to pay, for such success is that these fuzzy PID controllers are slightly more complicated than the conventional ones, in the sense that they have self-variable control gains in their linear structures. These variable gains are nonlinear functions of the tracking errors and changing rates of the error signals. The main contribution of these variable gains in improving the control performance is that they are self-tuned gains and can adapt to the rapid changes of the errors and the (changing) rates of the error signals caused by the time-delayed effects, nonlinearities, and uncertainties of the underlying system (plant, process).

A. Designing Type-2 Fuzzy PD Controller

The overall fuzzy PD set-point tracking control system is shown in Figure 6.14, where the process under control is a discrete-time system (or a discretized continuous-time system), and $r(nT)$ is the reference signal which can be a constant (set-point). The fuzzy PD controller inside the dashed box differs from the conventional digital PD controller (see Figure 6.7) in that there is an extra "fuzzy controller" in the path of the incremental control signal $\Delta u(nT)$. Moreover, a constant multiplication block has been changed from the sampling period T to an adjustable constant control gain K_u in order to provide the new controller with one more degree of freedom in the control process (but this is not necessary).

In this fuzzy PD controller, the "fuzzy controller" block is the key that improves the conventional digital PD controller's capabilities and performance.

To simplify the notation, let the adjustable control gains be

$$K_p = \tilde{K}_P \qquad \text{and} \qquad K_d = \tilde{K}_D$$

in Figure 6.14 (compared to Figure 6.7), and similarly let $K_i = \tilde{K}_I$ when the fuzzy PI controller is discussed later.

To illustrate how the "fuzzy controller" block works, first introduce a constant parameter (an adjustable scalar) $L > 0$, and decompose the plane by L as twenty input-combination regions (IC1-IC20) as shown in Figure 6.15, where the horizontal axis is the input signal $K_p d(nT)$, and the vertical axis the input signal $K_d v(nT)$, to the "fuzzy controller." Then, according to which region the input signals $(K_p d(nT), K_d v(nT))$ belong, the "fuzzy controller" block produces the following incremental outputs:

$$\Delta u(nT) = \frac{L[K_p d(nT) - K_d v(nT)]}{2(2L - K_p \, | \, d(nT) \, |)}, \qquad \text{in IC1, IC2, IC5, IC6,} \qquad (6.17)$$

$$= \frac{L[K_p d(nT) - K_d v(nT)]}{2(2L - K_d \, | \, d(nT) \, |)}, \qquad \text{in IC3, IC4, IC7, IC8,} \qquad (6.18)$$

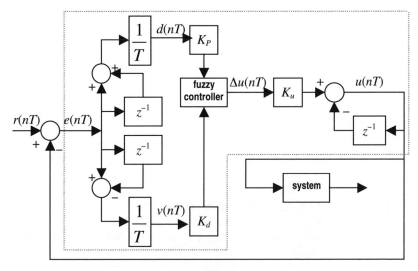

Figure 6.14   The fuzzy PD controller

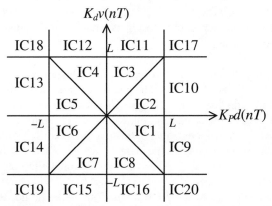

Figure 6.15 Input-combination regions of the "fuzzy controller"

$$= \frac{1}{2} [\, L - K_d v(nT)\,], \quad \text{in IC9, IC10,} \tag{6.19}$$

$$= \frac{1}{2} [\, -L + K_p d(nT)\,], \quad \text{in IC11, IC12,} \tag{6.20}$$

$$= \frac{1}{2} [\, -L - K_d v(nT)\,], \quad \text{in IC13, IC14,} \tag{6.21}$$

$$= \frac{1}{2} [L + K_p d(nT)], \quad \text{in IC15, IC16,} \quad (6.22)$$

$$= 0, \quad \text{in IC17, IC19,} \quad (6.23)$$

$$= -L, \quad \text{in IC18,} \quad (6.24)$$

$$= L, \quad \text{in IC20.} \quad (6.25)$$

Here, if the input signals $(K_p d(nT), K_d v(nT))$ belong to a boundary line, then either of the two neighboring regions can be used, since they will be the same (namely, all these control functions are continuous on the boundaries).

This completes the description for implementation and utilization of the fuzzy PD controller in a set-point tracking system, regardless of any knowledge about the given system (plant, process). The initial conditions for this control system are the following nature ones:

$$y(0) = 0, \quad \Delta u(0) = 0, \quad e(0) = r, \quad v(0) = 0. \quad (6.26)$$

Before discussing how to derive these formulas for the "fuzzy controller" block and in what sense this design is good, a few remarks on the above formulas are in order.

First, note that the above nine pieces of formulas are all conventional (crisp) analytical formulas, from which one does not see any fuzzy contents (membership functions, fuzzy logic IF-THEN rules, etc.). Therefore, the "fuzzy controller" block and the entire fuzzy PD controller shown in Figure 6.14 are conventional controllers: the overall control system works in the conventional manner despite the name "fuzzy." As a result, this fuzzy PD controller can be used to replace the conventional PD controller anywhere, and a control engineer can operate this fuzzy PD controller in a way completely analogous to the conventional one: what he needs to do is to tune the control gains and parameters, K_p, K_d, K_u, and L, without the need of knowing any fuzzy mathematics, fuzzy logic, and fuzzy control theory.

Second, the above nine pieces of formulas are continuously connected as a whole (on the boundaries between different regions shown in Figure 6.15). This can be verified by direct calculation of any two adjacent formulas on a boundary. Therefore, the control formula switching process does not have any jumps.

Third, since as usual all the control gains (K_p, K_d, and K_u) are positive real numbers, the above nine formulas can be computed from the two inputs $K_p d(nT)$ and $K_d v(nT)$, where one has

$$K_p \, | \, d(nT) \, | \, = \, | \, K_p v(nT) \, | \quad \text{and} \quad K_d \, | \, v(nT) \, | \, = \, | \, K_d v(nT) \, |.$$

Finally, but most importantly, it should be pointed out that the first formula (6.17) preserves the linear structure of the conventional PD controller (see formula (6.15)):

$$\Delta u(nT) \; = \; \left[\frac{LK_p}{2(2L - K_p \, | \, d(nT) \, |)} \right] d(nT) + \left[\frac{-LK_d}{2(2L - K_p \, | \, d(nT) \, |)} \right] v(nT),$$

except that the two control gains here are variable gains: they are nonlinear functions of the signal $d(nT) = \frac{1}{T} [e(nT) + e(nT{-}T)]$. Similarly, the second formula, (6.18), has the same linear structure but with two variable gains that are nonlinear functions of the signal $v(nT) = \frac{1}{T} [e(nT) - e(nT{-}T)]$.

It is very important to note that it is exactly these variable control gains that improve the performance of the controller: when the error signal $e(nT)$ increases, for instance, the signal $d(nT)$ increases, so that the control gain $L/2(2L{-}K_p|d(nT)|)$ also increases automatically. This means that the controller takes larger actions accordingly and, hence, has certain self-tuning and adaptive capabilities in the control process. Here, it is also important to note that $K_p|d(nT)|$ will not exceed the constant L in the denominator of the control gain according to the definition of the input-combination (IC) regions shown in Figure 6.15; otherwise, the formula will be switched to another one among the nine formulas.

Next, the design principle of the fuzzy PD controller described above is discussed, and detailed derivations to the resulting formulas (6.17)-(6.25) will be given.

First, recall that in a standard procedure, a fuzzy controller design consists of three components: (i) fuzzification, (ii) rule base establishment, and (iii) defuzzification. The intended design of the fuzzy PD controller follows this procedure.

In the fuzzification step, two inputs are employed: the error signal $e(nT)$ and the rate of change of the error signal $v(nT)$, with only one control output $u(nT)$ (to be fed to the system under control). The input to the fuzzy PD controller, namely, the "error" and the "rate" signals, have to be fuzzified before being fed into the controller. The membership functions for the two inputs (error and rate) and the output of the controller that is used in the design are shown in Figure 6.16, which are likely the simplest possible functions to use for this purpose. Both the error and the rate have two membership values: positive and negative, while the output has three (singleton functions): positive, negative, and zero. The constant $L > 0$ used in the definition of the membership functions is chosen by the designer according to the value ranges of the error, rate, and output, which is used as a tunable parameter but can also be fixed after being determined. Note that the constant L used in these three membership functions can be different in general (according to the physical meaning of the signals in the application), but one may let them be the same here in order to simplify the design.

Based on these membership functions, the fuzzy control rules that will be used are the following:

$R^{(1)}$ IF error = ep AND rate = vp THEN output = oz

$R^{(2)}$ IF error = ep AND rate = vn THEN output = op

$R^{(3)}$ IF error = en AND rate = vp THEN output = on

$R^{(4)}$ IF error = en AND rate = vn THEN output = oz

Here, "output" is the fuzzy control action $\Delta u(nT)$, "ep" means "error positive," "oz" means "output zero," etc. The "AND" is the logical AND defined by

$$\mu_A \text{ AND } \mu_B = \min\{ \mu_A, \mu_B \}$$

for any two membership values μ_A and μ_B on the fuzzy subsets A and B, respectively.

The reason for establishing the rules in such formulation can be understood in the same way as that described in Chapter 5, Section II, which is briefly repeated here for clarity. First, it is important to observe that since the error signal is defined to be $e = r - y$, where r is the reference (set-point) and y is the system output (see Figure 5.9 for the general situation in the continuous-time setting), one has $\dot{e} = \dot{r} - \dot{y} = -\dot{y}$ in the case that the set-point r is constant.

For Rule 1 ($R^{(1)}$): condition ep (the error is positive, $e > 0$) implies that $r > y$ (the system output is below the set-point) and condition vp ($\dot{e} > 0$) implies that $\dot{y} < 0$ (the system output is decreasing). In this case, the controller at the previous step is driving the system output, y, to move downward. Hence, the controller should turn around and drive the system output to move upward. It is very important to observe, however, that in this control law (see formula (6.15) and Figure 6.7):

$$u(nT) = -u(nT-T) + K_u \Delta u(nT), \tag{6.27}$$

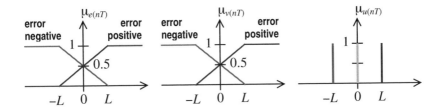

Figure 6.16 The membership functions of $e(nT)$, $v(nT)$, and $u(nT)$

there exists a minus sign in front of $u(nT–T)$, which will automatically perform the expected task. For this reason, one may set "output = oz" for the incremental control as the first rule, which means $\Delta u(nT) = 0$ at this step.

For Rule 2 ($R^{(2)}$): condition ep ($e > 0$) implies that $r > u$ and condition vn ($\dot{e} < 0$) implies that $\dot{y} > 0$. In this case, u is below r and the controller at the previous step is driving the system output, y, to move upward. Hence, the controller needs not to take any action. But, in control law (6.27), there is a minus sign with $u(nT–T)$ which will turn the control action to the opposite. To compensate for this, one should let Δu be positive, i.e., "output = op."

For Rule 3 ($R^{(3)}$): condition en ($e < 0$) implies that $r < y$ and condition vp ($\dot{e} > 0$) implies that $\dot{y} < 0$. In this case, y is over r and the controller at the previous step is driving the system output to move downward. Therefore, the controller needs not take any action. Similar to Rule 2, to compensate the minus sign in (6.27), let Δu be negative, i.e., "output = on."

For Rule 4 ($R^{(4)}$): condition en ($e < 0$) implies that $r < y$ and condition vn ($\dot{e} < 0$) implies that $\dot{y} > 0$. In this case, y is over r and the controller at the previous step is driving the system output to move upward. Hence, the controller should turn its output around to drive the system output to move downward. Because of the minus sign in (6.27), similar to Rule 1, the controller needs not take any action: "output = oz."

Here, it is remarked that one may try to let the controller take some action to speed up the system output in cases of Rules 1 and 4, but this will somewhat complicate the design of the controller.

In the defuzzification step, one may use the same logical AND as mentioned above, and the membership functions shown in Figure 6.16 for the error $e(nT)$, the rate $v(nT)$, and the output $\Delta u(nT)$ of the "fuzzy controller" block. Because one has two (positive and negative) membership values for the error and rate, the commonly used weighted average formula is used for defuzzification, leading to

$$\Delta u(nT) = \frac{\sum (\text{membership value of input} \times \text{corresponding value of output})}{\sum (\text{membership value of input})}$$

(6.28)

The defuzzification procedure and its corresponding results are now analyzed and summarized in the following.

First, observe from (6.17)-(6.25) that, instead of the error signal $e(nT)$, one will actually use the average error signal $d(nT)$. One will simply call $K_p d(nT)$ and $K_d r(nT)$ the "error signal" and the "rate signal," respectively. Note that using the average error will not alter the above reasoning for the control rules $(R^{(1)})$-$(R^{(4)})$, at least in principle.

Second, observe that the membership functions of the average error and rate signals decompose their value ranges into twenty adjacent input-combination (IC) regions, as shown in Figure 6.15. This figure is understood as follows.

Let the membership function of the error signal (given by the first picture of Figure 6.16) be put over the horizontal $K_p d(nT)$-axis in Figure 6.15, and put the membership function of the rate of change of the error signal (given by the second picture of Figure 6.16) over the vertical $K_d r(nT)$-axis in Figure 6.15. These two membership functions then overlap and form the third-dimensional picture (which is not shown in Figure 6.15) over the 2-D regions shown in Figure 6.15. When one looks at region IC1, for example, if one looks upward to the $K_p d(nT)$-axis, one can see the domain $[0,L]$ and the membership function (in the third dimension) over $[0,L]$ of the error signal; if one looks leftward to the $K_d r(nT)$-axis, one can see the domain $[-L,0]$ and the membership function (in the third dimension) over $[-L,0]$ of the rate of change of the error signal.

Then, consider the locations of the error $K_p d(nT)$ and the rate $K_d r(nT)$ in the region IC1 and IC2 (see Figure 6.15). Let us look at region IC1, for example, where one has ep > 0.5 > vp (see the first two pictures in Figure 6.16). Hence, the logical AND used in $(R^{(1)})$ leads to

{"error = ep AND rate = vp"} = min{ ep, vp } = vp,

so that Rule 1 $(R^{(1)})$ yields

$$R^{(1)}: \quad \begin{cases} \text{the selected input membership value is vp;} \\ \text{the corresponding output value is oz.} \end{cases}$$

Similarly, in region IC1, Rules 2-4, $(R^{(2)})$-$(R^{(4)})$, and the logical AND used in $(R^{(2)})$-$(R^{(4)})$ together yield

$$R^{(2)}: \quad \begin{cases} \text{the selected input membership value is vn;} \\ \text{the corresponding output value is op.} \end{cases}$$

$$R^{(3)}: \quad \begin{cases} \text{the selected input membership value is en;} \\ \text{the corresponding output value is on.} \end{cases}$$

$$R^{(4)}: \quad \begin{cases} \text{the selected input membership value is en;} \\ \text{the corresponding output value is oz.} \end{cases}$$

It can be verified that the above are true for the two regions IC1 and IC2. Thus, in regions IC1 and IC2, it follows from the defuzzification formula (6.28) that

$$\Delta u(nT) = \frac{\text{vp} \cdot \text{oz} + \text{vn} \cdot \text{op} + \text{en} \cdot \text{on} + \text{en} \cdot \text{oz}}{\text{vp} + \text{vn} + \text{en} + \text{en}}.$$

It is very important to note that if one follows the above procedure to work through the two cases, then it is found that both the last two cases give the same result of en (i.e., the two en in the above formula are not the misprint of en and ep!). To this end, by applying op = L, on = $-L$, oz = 0 (obtained from Figure 6.16), and the following straight line formulas from the geometry of the membership functions associated with Figure 6.15:

$$\text{ep} = \frac{K_p d(nT) + L}{2L}, \qquad \text{en} = \frac{-K_p d(nT) + L}{2L},$$

$$\text{vp} = \frac{K_d v(nT) + L}{2L}, \qquad \text{vn} = \frac{-K_d v(nT) + L}{2L},$$

one obtains

$$\Delta u(nT) = \frac{L}{2[2L - K_p d(nT)]} [K_p d(nT) - K_d v(nT)].$$

Here, it is noted that $d(nT) \geq 0$ in regions IC1 and IC2.

In the same way, one can verify that in regions IC5 and IC6,

$$\Delta u(nT) = \frac{L}{2[2L - K_p d(nT)]} [K_p d(nT) - K_d v(nT)],$$

where it should be noted that $d(nT) \leq 0$ in regions IC5 and IC6.

Therefore, by combining the above two formulas, one arrives at the following result for the four regions IC1, IC2, IC5, and IC6:

$$\Delta u(nT) = \frac{L}{2[2L - K_p |d(nT)|]} [K_p d(nT) - K_d v(nT)],$$

which is (6.17). Similarly, if $K_p d(nT)$ and $K_d v(nT)$ are located in the regions IC3, IC4, IC7, and IC8, one has

$$K_p |d(nT)| \leq K_d |v(nT)| \leq L,$$

and, in this case,

$$\Delta u(nT) = \frac{L}{2[2L - K_d |d(nT)|]} [K_p d(nT) - K_d v(nT)],$$

which is (6.18).

Finally, in the regions IC9-IC20, one obtains the corresponding formulas shown as in (6.19)-(6.25).

To this end, all the control rules and formulas for the fuzzy PID controller have been determined, with the control law (6.27) and the fuzzy control action $\Delta u(nT)$ calculated by (6.17)-(6.25) according to the different locations in Figure 6.15 of the error signal $K_p d(nT)$ and the rate of the change of the error signal $K_d v(nT)$. The initial conditions for the overall control system are the following natural values: for the fuzzy control action $\Delta u(0) = 0$, for the system output, $y(0) = 0$, for the original error and rate signals, $e(0) = r$ (the set-point) and $v(0) = 0$, respectively, as shown in (6.26).

Finally, it is remarked that in the steady-state situation, $|e(nT)| = 0$, so that $|v(nT)| = |d(nT)| = 0$ in the denominators of the coefficients of $\Delta u(nT)$. Thus, one has the steady-state relations between the conventional PD control gains K_d^c and K_p^c and the fuzzy PD control gain K_p and K_d as follows:

$$K_d^c = \frac{K_u K_d}{r} \qquad \text{and} \qquad K_p^c = \frac{K_u K_p}{r}. \qquad (6.29)$$

Example 6.5

In order to compare the above-derived fuzzy PD controller with the conventional one, first consider a first-order linear system with transfer function $H(s) = 1 / (s + 1)$, and reference (set-point) $r = 10.0$. For this sytem, the fuzzy controller parameters are: $T = 0.1$, $K_d = 0.5$, $K_p = 0.5$, $K_u = 1.0$, and $L = 361.0$. The control performance is shown in Figure 6.17.

Then consider a second-order linear system with transfer function $H(s) = 1 / (s^2 + 4s + 3)$, and the reference (set-point) $r = 10.0$. The system response, controlled by the fuzzy PD controller, is shown in Figure 6.18. The controller parameters are $T = 0.1$, $K_p = 0.51$, $K_d = 0.02$, $K_u = 0.232$, and $L = 1000.0$.

In these two cases, both conventional and fuzzy PD controllers work equally well.

To show one more case of the second-order linear system, consider one with the transfer function $H(s) = 1 / (s)(s + 100)$, which is only marginally stable. This time, let's make it more difficult by setting the reference as a ramp signal, $r(t) = t$. The controller parameters used are $T = 0.01$, $K_p = 25.0$, $K_d = -50.0$, $K_u = 0.5$, and $L = 3000.0$, and the tracking tolerance is 5% of the steady-state error. For this example, it is easy to verify that the steady-state error for the conventional PD controller is given analytically by $e_{ss}(t) = 100 / K_p^c$. Therefore, one has to use a high gain of $K_p^c = 2000.0$ (along with $K_d^c = 10.0$) to obtain specified performance (of 5% steady-state tracking error). The fuzzy control result is shown in Figure 6.19, where the solid curve is the set-point while the dashed curve is the system output. This result demonstrates the advantage of the fuzzy PD controller over the conventional one (the former uses very small control gains) even for a second-order linear process.

Next, consider a lower-order linear system with time-delay, with transfer function $H(s) = \dfrac{1}{(100s + 1)^2} e^{-3s}$. The comparison is shown in Figure 6.20. The conventional PD controller, with $K_p^c = 66.0$, $K_d^c = 25.0$, and $T = 0.1$, produces the solid curve. On the contrary, the fuzzy PD controller, with $T = 0.1$, $K_p = 49.3$, $K_d = 5.3$, $K_u = 0.8$, and $L = 19.0$, yields the dashed curve in the same figure.

Finally, the conventional and fuzzy PD controllers are compared using two nonlinear systems. The first one has the simple nonlinear model $\dot{y}(t) = 0.0001 |y(t)| + u(t)$. Two different references are used: the constant set-point $r = 1.0$ and the ramp signal $r(t) = t$. For the constant set-point case, the fuzzy PD controller has the parameters $T = 0.1$, $K_p = 19.5$, $K_d = 0.5$, $K_u = 0.1$, and $L = 20.0$. The result is shown in Figure 6.21(a). The conventional PD controller, on the other hand, cannot handle this nonlinear system no matter how one changes its two constant gains. One control performance is shown in Figure 6.21(b), with $K_p^c = 3.0$, $K_d^c = 0.1$, and $T = 0.1$. For the ramp signal reference case, the fuzzy PD controller produces a good tracking result with very small transient oscillation, shown in Figure 6.21(c), where $K_p = 19.0$, $K_d = 0.5$, $K_u = 0.1$, $L = 40.0$, and $T = 0.1$. However, although the conventional PD controller also performs well after a long transient period, as shown in Figure 6.21(d), its transient behavior is poorer (see Figure 6.21(e)), as compared to the fuzzy controller (Figure 6.21(c)).

The second nonlinear example is shown in Figure 6.22, where the system is described by $\dot{y}(t) = -y(t) + 0.5y^2(t) + u(t)$ with the constant set-point $r = 2.0$. The fuzzy controller is designed with $K_p = 41.9$, $K_d = 15.4$, $K_u = 0.1$, $L = 239.0$, and $T = 0.1$, which produces the tracking response shown in Figure 6.22. However, no matter how one adjusts the two constant gains of the conventional PD controller, it does not show any reasonable tracking results.

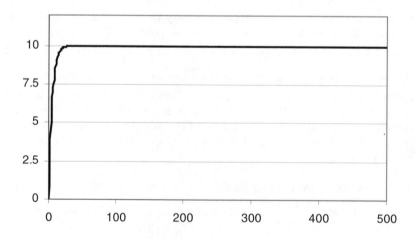

Figure 6.17 Output of a first-order linear fuzzy PD control system

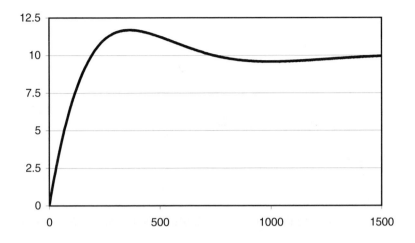

Figure 6.18 Output of a second-order linear fuzzy PD control system

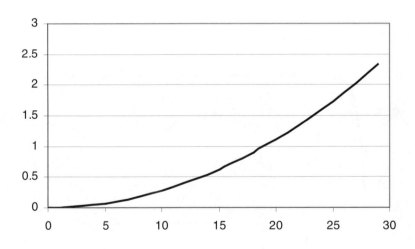

Figure 6.19 Output of a second-order linear fuzzy PD control system

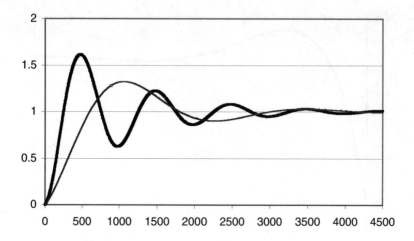

Figure 6.20 Comparison of outputs for a second-order linear fuzzy PD control system with a time delay

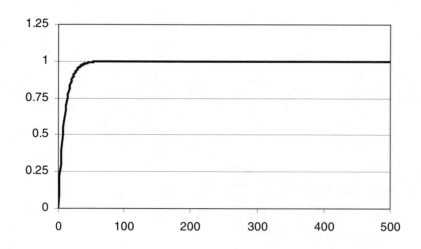

Figure 6.21(a) Output of a nonlinear fuzzy PD control system with a constant set-point

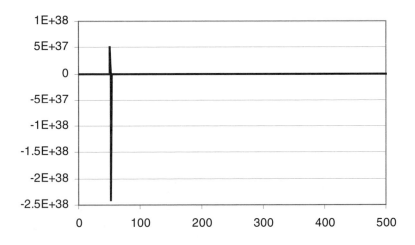

Figure 6.21(b) Output of a nonlinear conventional PD control system (with a constant set-point)

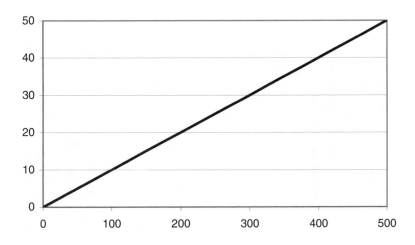

Figure 6.21(c) Output of a nonlinear fuzzy PD control system (with a ramp set-point)

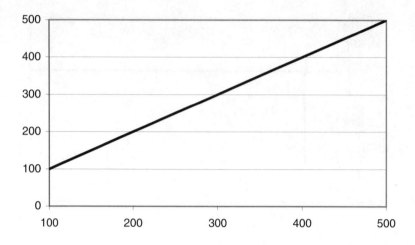

Figure 6.21(d) Steady-state output of a nonlinear conventional PD control system (with a ramp set-point)

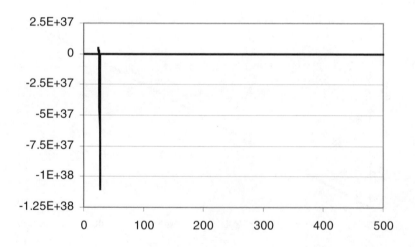

Figure 6.21(e) Transient output of a nonlinear conventional PD control system (with a ramp set-point)

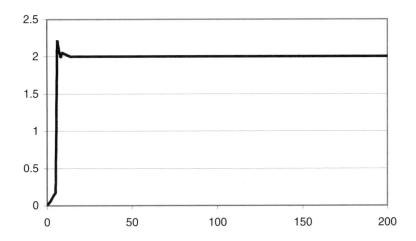

Figure 6.22 Output of a highly nonlinear PD control system

B. Designing Type-2 Fuzzy PI Controller

Design of the fuzzy PI controller is similar to that for the fuzzy PD controller, and will be introduced briefly in this subsection.

The overall PI controller system is shown in Figure 6.23 (see Figure 6.6). In Figure 6.23, let $K_p = \tilde{K}_P$ and $K_i = \tilde{K}_I$, as mentioned before. The "fuzzy controller" works in a way similar to that of the fuzzy PD controller. First, decompose the plane with a scalar $L > 0$ into twenty input-combination (IC) regions for the inputs $K_i e(nT)$ and $K_p v(nT)$, where $v(nT) = \frac{1}{T}[e(nT)-e(nT-T)]$, as shown in Figure 6.24. Here, $K_i e(nT)$ and $K_p v(nT)$ are called the error signal and the rate of change of error signal, respectively, for convenience. Then, according to the location of the inputs $(K_i e(nT), K_p v(nT))$ to the "fuzzy controller" block, the corresponding incremental control output are computed by the following formulas:

$$\Delta u(nT) \;=\; \frac{L[K_i e(nT) + K_p v(nT)]}{2(2L - K_i\,|e(nT)|)}, \qquad \text{in IC1, IC2, IC5, IC6,} \qquad (6.30)$$

$$=\; \frac{L[K_i e(nT) + K_p v(nT)]}{2(2L - K_p\,|e(nT)|)}, \qquad \text{in IC3, IC4, IC7, IC8,} \qquad (6.31)$$

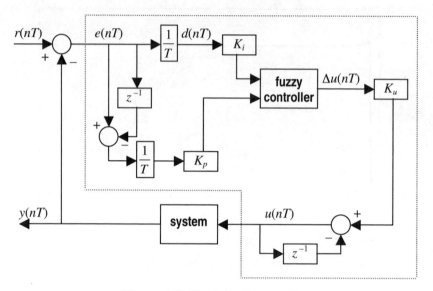

Figure 6.23 The fuzzy PI controller

$$= \frac{1}{2}[\, L + K_p v(nT)\,], \qquad\qquad \text{in IC9, IC10,} \qquad\qquad (6.32)$$

$$= \frac{1}{2}[\, L + K_i e(nT)\,], \qquad\qquad \text{in IC11, IC12,} \qquad\qquad (6.33)$$

$$= \frac{1}{2}[\, -L + K_p v(nT)\,], \qquad\qquad \text{in IC13, IC14,} \qquad\qquad (6.34)$$

$$= \frac{1}{2}[\, -L + K_i e(nT)\,], \qquad\qquad \text{in IC15, IC16,} \qquad\qquad (6.35)$$

$$= 0, \qquad\qquad\qquad\qquad\qquad \text{in IC18, IC20,} \qquad\qquad (6.36)$$

$$= -L, \qquad\qquad\qquad\qquad\qquad \text{in IC17,} \qquad\qquad\qquad (6.37)$$

$$= L, \qquad\qquad\qquad\qquad\qquad\quad \text{in IC19.} \qquad\qquad\qquad (6.38)$$

Then, the control action output of the fuzzy PI controller, which is the control input to the system (plant, process), is given by

$$u(nT) = u(nT{-}T) + K_u \Delta u(nT), \qquad\qquad (6.39)$$

where K_u is an adjustable constant control gain, which may be fixed to be $K_u = T$ to simplify the design (but one will then lose a degree of freedom in the tuning of the controller).

In a comparison of the fuzzy PI and PD control laws (6.39) and (6.27), one can find the main difference: there is a minus sign in front of $u(nT–T)$ in the fuzzy PD controller. This minus sign makes the rule base design quite different. Of course, their incremental controls $\Delta u(nT)$ are given by formulas (6.17)-(6.25) and (6.30)-(6.38), respectively, which are also different.

Similarly, the initial conditions for the fuzzy PI controller are the following natural ones:

$$y(0) = 0, \qquad \Delta u(0) = 0, \qquad e(0) = r, \qquad v(0) = 0. \qquad (6.40)$$

A verification of the above fuzzy PI controller formulas can be carried out in a step-by-step procedure by mimicking the fuzzy PD controller's design procedure given in the last subsection. This will be discussed in more detail in the next subsection within the PI+D controller.

It is remarked once again that the nine pieces of formulas (6.30)-(6.38) are conventional (crisp) formulas. Therefore, the "fuzzy controller" block as well as the entire fuzzy PI controller shown in Figure 6.23 are conventional controllers: the overall control system works in the conventional (crisp) manner despite the name "fuzzy." Therefore, this fuzzy PI controller can be used to replace the conventional digital PI controller anywhere. A control engineer can operate it without any knowledge of fuzzy mathematics, fuzzy logic, and fuzzy control theory: what he needs is to tune the control gains and parameters: K_p, K_i, K_u, and L.

Finally, it is remarked that computer simulations for comparison of the fuzzy and conventional PI controllers show similar results to those demonstrated in Example 6.5. More simulations will be shown in the next subsection.

C. Designing Type-2 Fuzzy PI+D Controller

The conventional analog PI+D controller is shown in Figure 6.2(d). Similar to the fuzzy PD controller design discussed above, one first discretizes it by applying the bilinear transform, then design the fuzzy PI and a fuzzy D controller separately, and finally combine them together as a whole in the closed-loop system.

In so doing, the design of the fuzzy PI controller is the same as that mentioned briefly above, but the design of the fuzzy D controller is rather different.

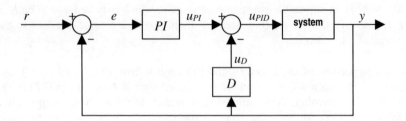

Figure 6.25 The conventional continuous-time PI+D control system

Start with the conventional analog PI+D control system shown in Figure 6.25. The output of the conventional analog PI controller in the frequency s-domain, as can be verified easily from Figure 6.3, is given by

$$U_{\text{PI}}(s) = \left[K_p^c + \frac{K_i^c}{s} \right] E(s),$$

where K_p^c and K_i^c are the proportional and integral gains, respectively, and $E(s)$ is the tracking error signal, after taking the Laplace transform (with zero initial conditions). This equation can be transformed into the discrete version by applying the bilinear transformation (5.8), which results in the following form:

$$U_{\text{PI}}(z) = \left[K_p^c - \frac{K_i^c T}{2} + \frac{K_i^c T}{1 - z^{-1}} \right] E(z).$$

Letting

$$K_p = K_p^c - \frac{K_i^c T}{2} \quad \text{and} \quad K_i = K_i^c\, T, \tag{6.41}$$

and then taking the inverse z-transform, one obtains

$$u_{\text{PI}}(nT) - u_{\text{PI}}(nT{-}T) = K_p[\, e(nT) - e(nT{-}T)\,] + K_i Te(nT).$$

Dividing this equation by T gives

$$\Delta u_{\text{PI}}(nT) = K_p v(nT) + K_i e(nT), \tag{6.42}$$

where

$$\Delta u_{\mathrm{PI}}(nT) = \frac{u_{\mathrm{PI}}(nT) - u_{\mathrm{PI}}(nT - T)}{T},$$

$$v(nT) = \frac{e(nT) - e(nT - T)}{T}.$$

It follows that

$$u_{\mathrm{PI}}(nT) = u_{\mathrm{PI}}(nT{-}T) + T\Delta u_{\mathrm{PI}}(nT).$$

In the design of the fuzzy PI controller to be discussed later, one may replace the coefficient T by a fuzzy control gain $K_{u,\mathrm{PI}}$, so that

$$u_{\mathrm{PI}}(nT) = u_{\mathrm{PI}}(nT{-}T) + K_{u,\mathrm{PI}}\Delta u_{\mathrm{PI}}(nT). \tag{6.43}$$

The D controller in the PI+D control system, as shown in Figure 6.25, has y as its input and u_{D} as its output. It is clear that

$$U_{\mathrm{D}}(s) = s\,K_d^{\,c}\,Y(s),$$

where $K_d^{\,c}$ is the control gain and $Y(s)$ is the output signal. Under the bilinear transformation, the above equation becomes

$$U_{\mathrm{D}}(z) = \frac{2}{T}\frac{z-1}{z+1}\,K_d^{\,c}\,Y(z),$$

or

$$U_{\mathrm{D}}(z) = K_d^{\,c}\,\frac{2}{T}\frac{1-z^{-1}}{1+z^{-1}}\,Y(z).$$

Consequently,

$$u_{\mathrm{D}}(nT) + u_{\mathrm{D}}(nT{-}T) = \frac{2K_d^{\,c}}{T}\,[\,y(nT) - y(nT{-}T)\,].$$

Then, dividing this equation by T yields

$$\Delta u_{\mathrm{D}}(nT) = K_d\Delta y(nT) \tag{6.44}$$

where

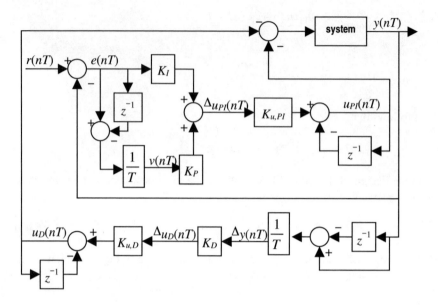

Figure 6.26 The conventional digital PI+D control system

$$\Delta u_D(nT) = \frac{u_D(nT) - u_D(nT - T)}{T}$$

is the incremental control output of the fuzzy D controller,

$$\Delta y(nT) = \frac{y(nT) - y(nT - T)}{T}$$

is the rate of change of the output y, and

$$K_d = \frac{2K_d^c}{T}.$$ (6.45)

As will be further described in the fuzzification step below, (6.44) will be modified by adding the signal $Ky_d(nT)$ to its right-hand side, where

$$y_d(nT) = y(nT) - r(nT) = - e(nT),$$

in order to obtain correct control rules. Thus, (6.44) becomes

$$\Delta u_D(nT) = K_d\Delta y(nT) + Ky_d(nT).$$

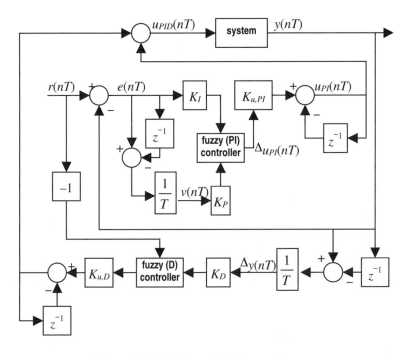

Figure 6.27 The fuzzy PI+D control system

Here, always set $K = 1$ in order to simplify the discussion in this section, which is not necessary in a real design.

Next, note from (6.44) that

$$\Delta u_\mathrm{D}(nT) = \frac{u_\mathrm{D}(nT) + u_\mathrm{D}(nT - T)}{T},$$

so that

$$u_\mathrm{D}(nT) = - u_\mathrm{D}(nT{-}T) + T\Delta u_\mathrm{D}(nT).$$

When $\Delta u_\mathrm{D}(nT)$ becomes a fuzzy control action later in the design, one uses $K_{u,\mathrm{D}}$ as this fuzzy control gain (which will be determined later in the design). Thus, one can rewrite the above formula as

$$u_\mathrm{D}(nT) = - u_\mathrm{D}(nT{-}T) + K_{u,\mathrm{D}}\Delta u_\mathrm{D}(nT). \qquad (6.46)$$

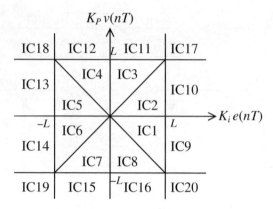

Figure 6.28 Input-combination regions of the "fuzzy (PI) controller"

Finally, the overall fuzzy PI+D control law can be obtained by algebraically summing the fuzzy PI control law (6.43) and fuzzy D law (6.46) together. The result is

$$u_{PID}(nT) = u_{PI}(nT) - u_D(nT),$$

or more precisely,

$$u_{PID}(nT) = u_{PI}(nT-T) + K_{u,PI}\Delta u_{PI}(nT) + u_D(nT-T) - K_{u,D}\Delta u_D(nT).$$

$$(6.47)$$

This equation will be referred to as the fuzzy PI+D control law below.

The overall conventional PI+D control system is shown in Figure 6.26. To this end, the fuzzy PI and fuzzy D controllers will be inserted into Figure 6.26, resulting in the configuration shown in Figure 6.27.

In summary, the fuzzy PI+D control system is implemented by Figure 6.27, in which there are five constant control gains that can be tuned: K_i, K_p, $K_{u,PI}$, K_D, and $K_{u,D}$, where one may set $K_{u,PI} = K_{u,D} = T$ to simplify the design. In this implementation, the two incremental control actions $\Delta u_{PI}(nT)$ and $\Delta u_D(nT)$ have the analytical formulas as shown below.

Similar to Figure 6.24, first divide the plane as twenty regions of the "fuzzy (PI) controller" input-combination (IC) for $(K_i e(nT), K_p v(nT))$, as shown in Figure 6.28. Then, the incremental control of the fuzzy PI controller is calculated by formulas (6.30)-(6.38), namely,

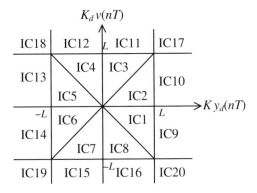

Figure 6.29   Input-combination regions of the "fuzzy (D) controller"

$$\Delta u_{\mathrm{PI}}(nT) = \frac{L[K_i e(nT) + K_p v(nT)]}{2(2L - K_i |e(nT)|)}, \quad \text{in IC1, IC2, IC5, IC6,} \tag{6.48}$$

$$= \frac{L[K_i e(nT) + K_p v(nT)]}{2(2L - K_p |e(nT)|)}, \quad \text{in IC3, IC4, IC7, IC8,} \tag{6.49}$$

$$= \frac{1}{2}[\, L + K_p v(nT)\,], \quad \text{in IC9, IC10,} \tag{6.50}$$

$$= \frac{1}{2}[\, L + K_i e(nT)\,], \quad \text{in IC11, IC12,} \tag{6.51}$$

$$= \frac{1}{2}[\, -L + K_p v(nT)\,], \quad \text{in IC13, IC14,} \tag{6.52}$$

$$= \frac{1}{2}[\, -L + K_i e(nT)\,], \quad \text{in IC15, IC16,} \tag{6.53}$$

$$= 0, \quad \text{in IC18, IC20,} \tag{6.54}$$

$$= -L, \quad \text{in IC17,} \tag{6.55}$$

$$= L, \quad \text{in IC19.} \tag{6.56}$$

For the fuzzy D controller, the twenty regions of the input-combination are shown in Figure 6.29. The incremental control of the fuzzy D controller is calculated by the following formulas:

$$\Delta u_{\mathrm{D}}(nT) \;=\; \frac{L[Ky_d(nT) + K_d\Delta y(nT)]}{2(2L - K\,|\,y_d(nT)\,|)}, \text{ in IC1, IC2, IC5, IC6,} \tag{6.57}$$

$$=\; \frac{L[Ky_d(nT) + K_d\Delta y(nT)]}{2(2L - K_d\,|\,y_d(nT)\,|)}, \text{ in IC3, IC4, IC7, IC8,} \tag{6.58}$$

$$=\; \frac{1}{2}[\,L - K_d\Delta y(nT)\,], \qquad \text{ in IC9, IC10,} \tag{6.59}$$

$$=\; \frac{1}{2}[\,-L + Ky_d(nT)\,], \qquad \text{ in IC11, IC12,} \tag{6.60}$$

$$=\; \frac{1}{2}[\,-L - K_d\Delta y(nT)\,], \qquad \text{ in IC13, IC14,} \tag{6.61}$$

$$=\; \frac{1}{2}[\,L + Ky_d(nT)\,], \qquad \text{ in IC15, IC16,} \tag{6.62}$$

$$=\; 0, \qquad\qquad\qquad\qquad \text{ in IC18, IC20,} \tag{6.63}$$

$$=\; -L, \qquad\qquad\qquad\qquad \text{ in IC17,} \tag{6.64}$$

$$=\; L, \qquad\qquad\qquad\qquad\; \text{ in IC19.} \tag{6.65}$$

Next, derivations of the above formulas of the fuzzy PI and D controllers are detailed.

First, one will fuzzify the PI and D components of the PI+D control system individually and then establish the desired fuzzy control rules for each of them taking into consideration the interconnected PI+D fuzzy control law given in equation (6.47). The input and output membership functions of the PI component are shown in Figures 6.30 and 6.31, respectively.

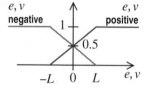

Figure 6.30 Input membership functions for the PI component

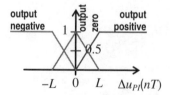

Figure 6.31 Output membership functions for the PI component

The fuzzy PI controller employs two inputs: the error signal $e(nT)$ and the rate of change of the error signal $v(nT)$. The fuzzy PI controller has a single output, $\Delta u_{PI}(nT)$, as shown in Figure 6.27, where the constant $L > 0$ is a tunable parameter (which can also be fixed after being determined).

Using these membership functions, the following control rules are established for the fuzzy PI controller:

$R^{(1)}$:   IF $e = en$ AND $v = vn$ THEN PI-output $= on$

$R^{(2)}$:   IF $e = en$ AND $v = vp$ THEN PI-output $= oz$

$R^{(3)}$:   IF $e = ep$ AND $v = vn$ THEN PI-output $= oz$

$R^{(4)}$:   IF $e = ep$ AND $v = vp$ THEN PI-output $= op$

In these rules, $e := r - y$ is the error, $v = \dot{e} = 0 - \dot{y} = -\dot{y}$ is the rate of change of the error, "PI-output" is the fuzzy PI control output $\Delta u_{PI}(nT)$, "ep" means "error positive," "op" means "output positive," etc. Also, AND is the logical AND operator defined as before.

If one looks at the fuzzy control law for the D component, equation (6.44), the only information it contains that is relevant to the output performance is $\Delta y(nT)$. Based on this signal alone, it is impossible to come up with a useful fuzzy control law. One therefore needs to look for another control signal that can be used in conjunction with $\Delta y(nT)$, to provide information about the output (above or below the reference signal). For this purpose, a logical and natural choice is the negative error signal

$$y_d(nT) = -e(nT). \tag{6.66}$$

Here, it is important to observe that y_d positive (resp. negative) means that the system output y is above (resp. below) the reference r. This y_d control signal is implemented as shown by the path with the -1 block in Figure 6.27 (compared with Figure 6.26).

The input and output membership functions for the fuzzy D controller are shown in Figures 6.32 and 6.33, respectively.

Similarly, from the membership functions of the fuzzy D controller, the following control rules are used for the D component:

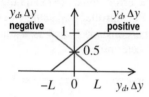

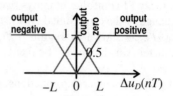

Figure 6.32   Input membership functions for the D component

Figure 6.33 Output membership functions for the D component

$R^{(5)}$:   IF $y_d = y_d p$ AND $\Delta y = \Delta yp$ THEN D-output = oz

$R^{(6)}$:   IF $y_d = y_d p$ AND $\Delta y = \Delta yn$ THEN D-output = op

$R^{(7)}$:   IF $y_d = y_d n$ AND $\Delta y = \Delta yp$ THEN D-output = on

$R^{(8)}$:   IF $y_d = y_d n$ AND $\Delta y = \Delta yn$ THEN D-output = oz

In the above rules, "D-output" is the fuzzy D control output $\Delta u_D(nT)$, and the other terms are defined similar to the PI components.

These eight rules altogether yield the control actions for the fuzzy PI+D control law.

Formulation of these rules can be understood as follows. For Rule 1 ($R^{(1)}$): if one looks at this rule for the PI controller, condition "en" (error is negative) implies that the system output, y, is above the set-point, and "vn" (rate of error is negative) implies $\dot{y} > 0$ (meaning that the controller at the previous step is driving the system output to move upward). Since the $\Delta u_{PI}(nT)$ component of formula (6.47) contains more control terms with gain parameters than the D controller, one may set this term to be negative and set the $\Delta u_D(nT)$ component to be zero. Thus, the combined control action will drive the system output to move downward by Rules 1 and 5 of both controllers.

Note that one could have set the output of the D controller to be positive in order to drive the output to move downward faster. However, having both controllers' outputs being nonzero at the same time will complicate the design of the controller. Computer simulations have demonstrated that the simple design described above performs sufficiently well in all numerical examples tested, so more sophisticated design is not discussed here.

Similarly, for Rule 2 ($R^{(2)}$), since the output is above the set-point and is moving downward, one can set the "larger" component of formula (6.47), namely, the term $\Delta u_{PI}(nT)$, to be zero, and set the "smaller" component $\Delta u_D(nT)$ to be positive. Thus, the combined controller will tend to drive the system output to move downward faster by the combined action of these two rules.

Rules 3 and 4 are similarly determined.

In the defuzzification step, for both fuzzy PI and D controllers, the same weighted average formula is employed to defuzzify the incremental control of the fuzzy control law:

$$\Delta u(nT) = \frac{\sum (\text{membership value of input} \times \text{corresponding value of output})}{\sum (\text{membership value of input})}$$

(6.67)

For the fuzzy PI controller, the value ranges of the two inputs, the error, and the rate of change of the error are actually decomposed into twenty adjacent input-combination (IC) regions, as shown in Figure 6.28. This figure is understood as follows: One may put the membership function of the error signal (given by the curves for e in Figure 6.30) over the horizontal $K_i e(nT)$-axis on Figure 6.28, and put the membership function of the rate of change of the error signal (given by the same curves in Figure 6.30 for v) over the vertical $K_p v(nT)$-axis on Figure 6.28. These two membership functions then overlap and form the third-dimensional picture (which is not shown in Figure 6.28) over the two-dimensional regions shown in Figure 6.28. When one looks at region IC1, for example, if one looks upward to the $K_i e(nT)$-axis, one can see the domain $[0,L]$ and the membership function (in the third dimension) over $[0,L]$ of the error signal; if one looks leftward to the $K_p v(nT)$-axis, one can see the domain $[-L,0]$ and the membership function (in the third dimension) over $[-L,0]$ of the rate of change of the error signal. This situation is completely analogous to the analysis of the fuzzy PD controller studied before.

The control rules for the fuzzy PI controller ($R^{(1)}$-$R^{(4)}$), with membership functions and IC regions together, are used to evaluate appropriate fuzzy control formulas for each region.

In doing so, consider the locations of the error $K_i e(nT)$ and the rate $K_p v(nT)$ in the regions IC1 and IC2 (see Figure 6.28). Look at region IC1, for example, where one has $e > 0.5 > v(nT)$ (see Figure 6.30). Hence, the logical AND used in ($R^{(1)}$) leads to

$$\{\text{"error} = e \text{ AND rate} = v\text{"}\} = \min\{ e, v \} = e,$$

so that Rule 1 ($R^{(1)}$) yields

$$R^{(1)}: \quad \begin{cases} \text{the selected input membership value is } en; \\ \text{the corresponding output value is } on. \end{cases}$$

Similarly, in region IC1, Rules 2-4, $(R^{(2)})$-$(R^{(4)})$, are obtained as

$$R^{(2)}: \quad \begin{cases} \text{the selected input membership value is } en; \\ \text{the corresponding output value is } oz. \end{cases}$$

$$R^{(3)}: \quad \begin{cases} \text{the selected input membership value is } vn; \\ \text{the corresponding output value is } oz. \end{cases}$$

$$R^{(4)}: \quad \begin{cases} \text{the selected input membership value is } vp; \\ \text{the corresponding output value is } op. \end{cases}$$

It can be verified that the above are true for the two regions IC1 and IC2. Thus, in regions IC1 and IC2, it follows from the defuzzification formula (6.67) that

$$\Delta u(nT) = \frac{en \times on + en \times oz + vn \times oz + vp \times op}{en + en + vn + vp}.$$

It is very important to note that if one follows the above procedure to work through the two cases, then one will find that both the last two cases give the same result of en (i.e., the two en in the above formula are not the misprint of en and ep!). To this end, by applying $op = L$, $on = -L$, $oz = 0$ (obtained from Figure 6.31), and the following straight line formulas from the geometry of the membership functions associated with Figure 6.28:

$$ep = \frac{K_i e(nT) + L}{2L}, \qquad en = \frac{-K_i e(nT) + L}{2L},$$

$$vp = \frac{K_p v(nT) + L}{2L}, \qquad vn = \frac{-K_p v(nT) + L}{2L},$$

one obtains

$$\Delta u_{PI}(nT) = \frac{L[K_i e(nT) + K_p v(nT)]}{2[2L - K_i e(nT)]}.$$

Here, note that $e(nT) \geq 0$ in regions IC1 and IC2. In the same way, one can verify that in regions IC5 and IC6,

$$\Delta u_{\text{PI}}(nT) = \frac{L[K_i e(nT) + K_p v(nT)]}{2[2L - K_i e(nT)]} \,,$$

where it should be noted that $e(nT) \leq 0$ in regions IC5 and IC6. Thus, by combining the above two formulas, one arrives at the following control formula for the four regions IC1, IC2, IC5, and IC6:

$$\Delta u_{\text{PI}}(nT) = \frac{L[K_i e(nT) + K_p v(nT)]}{2[2L - K_i e(nT)]} \,.$$

Working through all regions in the same way, one obtains the PI control formulas (6.48)-(6.56) for the twenty IC regions.

Similarly, defuzzification of the fuzzy D controller follows the same procedure as described above for the PI component, except that the input signals in this case are different. The IC combinations of these two inputs are decomposed into twenty similar regions, as shown in Figure 6.29.

Similarly, by applying the value $op = L$, $on = -L$, $oz = 0$, and the following straight line formulas obtained from the geometry of Figure 6.33:

$$y_dp = \frac{Ky_d(nT) + L}{2L} \,, \qquad\qquad y_dn = \frac{-Ky_d(nT) + L}{2L} \,,$$

$$\Delta yp = \frac{K_d \Delta y(nT) + L}{2L} \,, \qquad\qquad \Delta yn = \frac{-K_d \Delta y(nT) + L}{2L} \,,$$

one obtains the D control formulas (6.57)-(6.65) for the twenty IC regions.

Note that the constant K that multiplies the signal $y_d(nT)$ is used as a parameter for generality here, and in the derivation of $\Delta u_D(nT)$. Although it could be used as a control gain, its value is permanently set to one throughout the computer simulations shown in the next example.

Example 6.6

First, apply the fuzzy PI+D controller to a lower-order linear system, to see how well it performs for such simple cases. Recall that the conventional PI+D controller is designed for linear systems, for which it works very well. It will be seen that the fuzzy PI+D controller is as good as, if not better than, the conventional one for such lower-order linear systems.

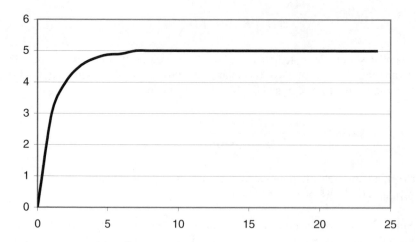

Figure 6.34 Output of a first-order linear fuzzy PI+D control system

The first example is a first-order linear system, with transfer function $H(s) = \dfrac{1}{s+1}$ and with controller parameters $T = 0.1$, $K = 1.0$, $K_p = 1.2$, $K_i = 1.0$, $K_{u,\mathrm{PI}} = 0.2$, $K_{u,\mathrm{D}} = 0.01$, $L = 360.0$. The set-point is $r = 5.0$. The response of the fuzzy PI+D controller for a step input is shown in Figure 6.34.

The second example is a second-order linear system with transfer function $H(s) = \dfrac{1}{s^2 + 4s + 3}$, where the controller parameters are $T = 0.01$, $K = 1.0$, $K_p = 8.0$, $K_d = 0.01$, $K_i = 10.0$, $K_{u,\mathrm{PI}} = 0.2$, $K_{u,\mathrm{D}} = 0.01$, $L = 10.0$, and the set-point is $r = 5.0$. The response of the fuzzy PI+D controller is shown in Figure 6.35.

Finally, compare the performance of both the conventional and the fuzzy PI+D controllers, using two nonlinear systems. Although the fuzzy PI+D controller has the same linear structure as the conventional PI+D controller, the fuzzy PI+D gains are nonlinear with self-tuning capability and, therefore, have better performance in this simulation and in general.

The first nonlinear system has the following simple model: $\dot{y}(t) = 0.0001\,|y(t)| + u_{\mathrm{PID}}(t)$, with fuzzy PI+D parameters $T = 0.1$, $K = 1.0$, $K_p = 1.5$, $K_d = 0.1$, $K_i = 2.0$, $K_{u,\mathrm{PI}} = 0.11$, $K_{u,\mathrm{D}} = 1.0$, $L = 45.0$, and the set-point is $r = 5.0$. The result is shown in Figure 6.36. On the contrary, the conventional PI+D controller is unable to track the set-point, no matter how one changes its parameters. A typical response of the conventional PI+D controller is shown in Figure 6.37, with parameters $T = 0.1$, $K_p^c = 19.5$, $K_i^c = 1.0$, for $r = 5.0$.

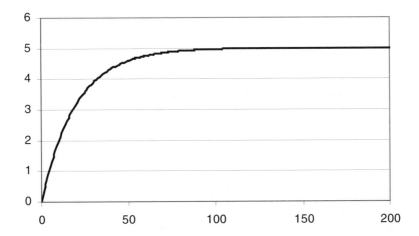

Figure 6.35 Output of a second-order linear fuzzy PI+D control system

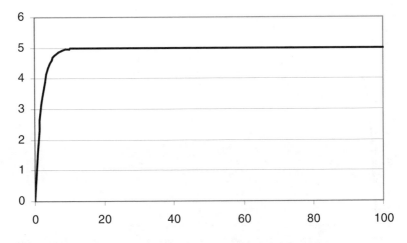

Figure 6.36 Output of a nonlinear fuzzy PI+D control system

The second nonlinear system used in the simulation is $\dot{y}(t) = y(t) + \sqrt{y(t)} + u_{\mathrm{PID}}(t)$. In this case, the fuzzy PI+D parameters are $T = 0.1$, $K = 1.0$, $K_p = 2.0$, $K_d = 1.942$, $K_i = 1.0$, $K_{u,\mathrm{PI}} = 0.1$, $K_{u,\mathrm{D}} = 0.27$, $L = 350.0$, and the set-point $r = 5.0$. The response of this fuzzy PI+D control system to a step input is shown in Figure 6.38. The conventional PI+D controller, however, cannot yield any reasonable response, no matter how one adjusts its gains. A typical response of the conventional PI+D controller (with parameters $T = 0.1$, $K_p^c = 2.0$, $K_i^c =$

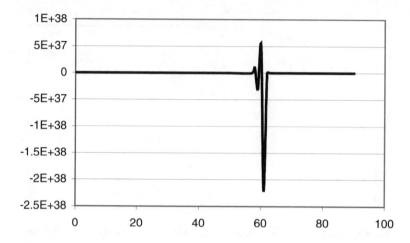

Figure 6.37 Output of a nonlinear conventional PI+D control system

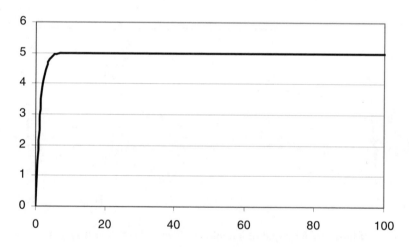

Figure 6.38 Output of a nonlinear fuzzy PI+D control system

1.0, and $r = 5.0$) is shown in Figure 6.39. It is remarked that this result is due to the fact that the conventional PI+D controller usually has difficulties in controlling higher-order and time-delayed linear systems as well as nonlinear systems, because they are designed only for lower-order linear systems, for which they can work very well as has been widely experienced.

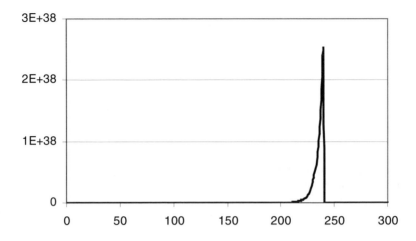

Figure 6.39 Output of a nonlinear conventional PI+D control system

IV*. FUZZY PID CONTROLLERS: STABILITY ANALYSIS

In this section, the stability of the fuzzy PD, PI, and PI+D control systems designed and studied in the last section is analyzed.

Recall the Lyapunov asymptotic stability for fuzzy modeling discussed in Chapter 4. In order to show another important type of stability, namely, the bounded-input bounded-output (BIBO) stability of a control system, the Small Gain Theorem is introduced in this section and discuss the BIBO stability of the fuzzy PI, PD, and PI+D control systems. First, note that differing from the Lyapunov asymptotic stability, which is usually local (in a neighborhood of an equilibrium point), the BIBO stability is global and is particularly suitable for nonlinear systems described by input-output maps. It should also be noted that both the Lyapunov asymptotic stability and the BIBO stability analyses provide conservative sufficient conditions, especially for nonlinear systems. From a theoretical point of view, the larger the region of stability can be found, the better the result is. However, from the design point of view, relatively conservative stability region is actually safer and more reliable in applications. BIBO stability theory turns out to be appropriate for this purpose.

Definition 6.1. A (linear or nonlinear) control system is said to be *bounded-input bounded-output (BIBO) stable* if a bounded control input to the system always produces a bounded output through the system.

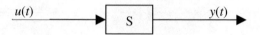

Figure 6.40 Input-output relation of a system

Here, the boundedness is defined in the norm (l_2, l_∞, etc.) of the function space, which is considered in the design.

Let S denote a (linear or nonlinear) system. S may be considered as a *mapping* which maps a control input, $u(t)$, to the corresponding system output $y(t)$, as shown in Figure 6.40, where

$$S: u(t) \to y(t) \qquad \text{or} \qquad y(t) = S\{u(t)\}.$$

Recall the standard L_p-spaces of signals:

$$1 \le p < \infty: \qquad L_p = \{f(t) | \int_0^\infty | f(t) |^p \, dt < \infty \},$$

$$p = \infty: \qquad L_p = \{f(t) | \ ess \ \sup_{0 \le t < \infty} |f(t)| < \infty \},$$

where "*ess*" means "essential," namely, the supirum holds except over a set of measure zero. For piecewise continuous signals, essential supirum and supirum are the same, so "*ess*" can be dropped from the above.

A. BIBO Stability and the Small Gain Theorem

Consider a nonlinear (including linear) feedback system shown in Figure 6.41, where for simplicity it is assumed that all signals u, e, y_1, u_2, e_2, $y_2 \in R^n$. It is clear from Figure 6.41 that

$$\begin{cases} e_1 = u_1 - S_2(e_2), \\ e_2 = u_2 - S_1(e_1), \end{cases} \tag{6.68}$$

or, equivalently,

$$\begin{cases} u_1 = e_1 + S_2(e_2), \\ u_2 = e_2 + S_1(e_1), \end{cases} \tag{6.69}$$

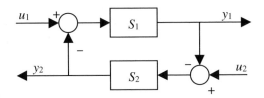

Figure 6.41 A nonlinear feedback system

For this system, the following so-called Small Gain Theorem holds, which gives sufficient conditions under which a "bounded input" yields a "bounded output," where the norm $\| \cdot \|$ is the standard Euclidean norm ("length" of a vector).

Theorem 6.1. Consider the nonlinear feedback system shown in Figure 6.41, which is described by the relationship (6.68)-(6.69). Suppose that there exist constants L_1, L_2, M_1, M_2, with $L_1 L_2 < 1$, such that

$$\begin{cases} \| S_1(e_1) \| \leq M_1 + L_1 \| e_1 \|, \\ \| S_2(e_2) \| \leq M_2 + L_2 \| e_2 \|. \end{cases} \tag{6.70}$$

Then,

$$\begin{cases} \| e_1 \| \leq (1 - L_1 L_2)^{-1} (\| u_1 \| + L_2 \| u_2 \| + M_2 + L_2 M_1), \\ \| e_2 \| \leq (1 - L_1 L_2)^{-1} (\| u_2 \| + L_1 \| u_1 \| + M_1 + L_1 M_2). \end{cases} \tag{6.71}$$

Proof. Since

$$e_1 = u_1 - S_2(e_2)$$

one has

$$\| e_1 \| \leq \| u_1 \| + \| S_2(e_2) \| \leq \| u_1 \| + M_2 + L_2 \| e_2 \|.$$

Similarly,

$$\| e_2 \| \leq \| u_2 \| + M_1 + L_1 \| e_1 \|.$$

Combining these two inequalities yields

$$\| e_1 \| \leq L_1 L_2 \| e_1 \| + \| u_1 \| + L_2 \| u_2 \| + M_2 + L_2 M_1,$$

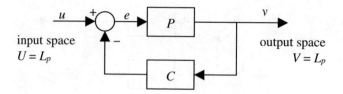

Figure 6.42 A feedback control system

or, using the fact $L_1L_2 < 1$,

$$\| e_1 \| \le (1-L_1L_2)^{-1}(\| u_1 \| + L_2 \| u_2 \| + M_2 + L_2M_1).$$

The rest of the theorem follows immediately.

It is clear that the Small Gain Theorem is applicable to both continuous-time and discrete-time systems, and to both SISO and MIMO systems. Hence, although its statement and proof are quite simple, it is very useful.

There is an interesting relation between the BIBO stability and the Lyapunov asymptotic stability. It is clear that the asymptotic stability generally implies the BIBO stability, but the reverse can also be true under some conditions.

Consider a nonlinear system described by the following first-order vector-valued ordinary differential equation:

$$\begin{cases} \dot{x}(t) = Ax(t) - f(x(t),t), \\ x(0) = x_0, \end{cases} \qquad (6.72)$$

with an equilibrium solution $\bar{x}(t) = 0$, where A is an $n \times n$ constant matrix whose eigenvalues are assumed to have negative real parts, and $f{:}R^n{\times}R^1{\rightarrow}R^n$ is a real vector-valued integrable nonlinear function of $t \in [0,\infty)$. By adding and then subtracting the term $Ax(t)$, a general nonlinear system can always be written in this form. Let

$$\begin{cases} x(t) = u(t) - \int_0^t e^{(t-\tau)} y(\tau)d\tau, \\ y(x) = f(x(t),t), \end{cases} \qquad (6.73)$$

with $u(t) = e^{At}x_0$. Then one can implement system (6.73) by a feedback configuration as depicted in Figure 6.42, where the error signal $e(t) = x(t)$, the plant $P(\cdot)(t) = f(\cdot,t)$, and the compensator $C(\cdot)(t) = \int_0^t e^{(t-\tau)A}(\cdot)(\tau)d\tau$.

Theorem 6.2. Consider the nonlinear system (6.73) and its associate feedback configuration shown in Figure 6.42. Suppose that $U = V = L_p([0,\infty),R^n)$, where $1 \le p \le \infty$. Then, if the feedback system shown in Figure 6.42 is BIBO stable, then it is also asymptotically stable.

Proof. Since all eigenvalues of the constant matrix A have negative real parts, one has

$$| e^{tA} x_0 | \le Me^{-\alpha t}$$

for some constants $0 < \alpha$, $M < \infty$ for all $t \in [0,\infty)$, so that $|u(t)| = | e^{tA} x_0 | \to 0$ as $t \to \infty$. Hence, in view of the first equation defined above, i.e., $x(t) = u(t) - \int_0^t e^{(t-\tau)A} y(\tau)d\tau$, if one can prove that

$$v(t) := \int_0^t e^{(t-\tau)A} y(\tau)d\tau \to 0 \quad \text{as} \quad t \to \infty,$$

then it will follow that

$$| x(t) | = | u(t) - v(t) | \to 0 \quad \text{as} \quad t \to \infty.$$

To do so, write

$$v(t) = \int_0^{t/2} e^{(t-\tau)A} y(\tau)d\tau + \int_{t/2}^t e^{(t-\tau)A} y(\tau)d\tau$$

$$= \int_{t/2}^t e^{\tau A} y(t-\tau)d\tau + \int_{t/2}^t e^{(t-\tau)A} y(\tau)d\tau.$$

Then, by the Hölder inequality, one has

$$| v(t) | \le \left| \int_{t/2}^t e^{\tau A} y(t-\tau)d\tau \right| + \left| \int_{t/2}^t e^{(t-\tau)A} y(\tau)d\tau \right|$$

$$\le \left[\int_{t/2}^t | e^{\tau A} |^q d\tau \right]^{1/q} \left[\int_{t/2}^t | y(t-\tau) |^p d\tau \right]^{1/p}$$

$$+ \left[\int_{t/2}^t | e^{(t-\tau)A} |^q d\tau \right]^{1/q} \left[\int_{t/2}^t | y(\tau) |^p d\tau \right]^{1/p}$$

$$\le \left[\int_{t/2}^{\infty} |e^{\tau A}|^q \, d\tau \right]^{1/q} \left[\int_0^{\infty} |y(t-\tau)|^p \, d\tau \right]^{1/p}$$

$$+ \left[\int_{t/2}^{\infty} |e^{(t-\tau)A}|^q \, d\tau \right]^{1/q} \left[\int_0^{\infty} |y(\tau)|^p \, d\tau \right]^{1/p}.$$

Since all eigenvalues of A have negative real parts and since the feedback system is BIBO stable from U to V, so that $y \in V = L_p([0,\infty),R^n)$, one has

$$\lim_{t \to \infty} \int_{t/2}^{\infty} |e^{\tau A}|^q \, d\tau = 0$$

and

$$\lim_{t \to \infty} \int_{t/2}^{\infty} |y(\tau)|^p \, d\tau = 0.$$

Therefore, it follows that $|v(t)| \to \infty$ as $t \to \infty$, completing the proof of the theorem.

B. BIBO Stability of Fuzzy PD Control Systems

The BIBO stability of the fuzzy PD, PI, and PI+D control systems is now discussed. First, we discuss in detail the fuzzy PD system. Let's return to the fuzzy PD control system described in Figure 6.14. Consider the general case where the system under control is nonlinear, which will be denoted by N. Suppose that the fuzzy control law (6.27) together with the incremental control formula (6.17) are used, and let the reference signal be $r = r(nT)$ for generality. By defining

$$\begin{cases} e_1(nT) &= e(nT), \\ e_2(nT) &= u(nT), \\ u_1(nT) &= r(nT), \\ u_2(nT) &= -u(nT-T), \\ S_1(e_1(nT)) &= K_u \Delta u(nT), \\ S_2(e_2(nT)) &= N(e_2(nT)). \end{cases} \tag{6.74}$$

it is easy to see that an equivalent closed-loop control system as shown in Figure 6.43 is

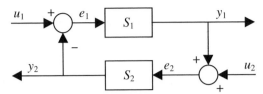

Figure 6.43 An equivalent closed-loop control system

$$\begin{cases} u_1(nT) & = & r(nT) = e(nT) + N(u(nT)) \\ & = & e_1(nT) + S_2(e_2(nT)), \\ u_2(nT) & = & -u(nT-T) = u(nT) - K_u\Delta u(nT) \\ & = & e_2(nT) + S_1(e_1(nT)). \end{cases} \tag{6.75}$$

Observe that when $e(nT)$ and $r(nT)$ are in the regions IC1, IC2, IC5, IC6, one has formula (6.17), so that

$$\| S_1(e_1(nT)) \| \leq \frac{K_u L}{2(2L - K_p M_e)}\left[\frac{|K_p - K_d|}{T}|e_1(nT)| + |e(nT-T)|\right]$$

$$= \frac{K_u L |K_p - K_d|}{2(2L - K_p M_e)} M_e + \frac{K_u L |K_p - K_d|}{2(2L - K_p M_e)}|e_1(nT)|, \tag{6.76}$$

and

$$\| S_2(e_2(nT)) \| \leq \| N \| \cdot | e_2(nT) |, \tag{6.77}$$

where $\| N \|$ is the operator norm of the given $N(\cdot)$, or the gain of the given nonlinear system, defined as usual by

$$\| N \| := \sup_{v_1 \neq v_2, n \geq 0} \frac{|N(v_1(nT)) - N(v_2(nT))|}{|v_1(nT) - v_2(nT)|} \tag{6.78}$$

over a set of admissible control signals that have any meaningful function norms, and M_e is defined by

$$M_e := \sup_{n \geq 1} |d(nT)| = \sup_{n \geq 1} \frac{2}{T}|e(nT)|. \tag{6.79}$$

To this end, an application of the Small Gain Theorem (Theorem 6.1) yields the following sufficient condition for the BIBO stability of the nonlinear fuzzy PD control systems:

$$\frac{K_u L |K_p - K_d|}{2T(2L - K_p M_e)} \, \|N\| < 1. \tag{6.80}$$

When $e(nT)$ and $r(nT)$ are in the regions IC3, IC4, IC7, IC8, one can similarly obtain a sufficient stability condition as follows:

$$\frac{K_u L |K_p - K_d|}{2T(2L - K_p M_r)} \, \|N\| < 1, \tag{6.81}$$

where

$$M_r := \sup_{n \geq 1} |r(nT)| = \sup_{n \geq 1} \frac{1}{T} |e(nT) - e(nT{-}T)| \leq M_e. \tag{6.82}$$

When $e(nT)$ and $r(nT)$ are in the rest of the regions, from IC9-IC20, the other incremental control formulas (6.19)-(6.25) are used. In these cases, the stability conditions are found to be

$$\begin{cases} \dfrac{K_u K_p}{2T} \, \|N\| < 1, \\[3mm] \dfrac{K_u K_d}{2T} \, \|N\| < 1, \\[3mm] \|N\| \text{ is bounded.} \end{cases} \tag{6.83}$$

By combining all the above conditions together, and noting that in IC1-IC8, $K_p M_e \leq L$, and that $K_p > 0$ and $K_d > 0$, one arrives at the following result for the stability of the nonlinear fuzzy PD control systems.

Theorem 6.3. A sufficient condition for the nonlinear fuzzy PD control systems to be BIBO stable is that the given nonlinear system has a bounded norm (gain) $\|N\| < \infty$ and the parameters of the fuzzy PD controller, K_p, K_d, and K_u, satisfy

$$\frac{\gamma K_m K_u}{2TL} \, \|N\| < 1, \tag{6.84}$$

where

$$\gamma = \max\{ 1, L \} \qquad \text{and} \qquad K_m = \max\{ K_p, K_d \}.$$

It is remarked that this theorem provides a useful criterion for the design of the nonlinear fuzzy PD controller when a nonlinear process N is given. One may first choose $K_u = T$ and then find a value of K_d (or K_p) for the tracking purpose

$$e(nT) = y(nT) - r(nT) \rightarrow 0 \quad \text{as} \quad n \rightarrow \infty. \qquad (6.85)$$

Finally, one may determine K_d (or K_p) among all possible choices such that the inequality (6.84) is satisfied.

C. BIBO Stability of Fuzzy PI Control Systems

The analysis of the BIBO stability condition for fuzzy PI control systems is similar to that for fuzzy PD control systems discussed in the last subsection.

Again, consider the general case where the system under control, N, is nonlinear. Start with the configuration shown in Figure 6.23 with the control law (6.39) and (6.40). By defining

$$\begin{cases} e_1(nT) &= e(nT), \\ e_2(nT) &= u(nT), \\ u_1(nT) &= r(nT), \\ u_2(nT) &= -u(nT-T), \\ S_1(e_1(nT)) &= K_u \Delta u(nT), \\ S_2(e_2(nT)) &= N(e_2(nT)), \end{cases} \qquad (6.86)$$

it is easy to see that one obtains an equivalent closed-loop control system as shown in Figure 6.43 where, differing from the fuzzy PD control system, one has

$$\begin{cases} u_1(nT) &= r(nT) = e(nT) + N(u(nT)) \\ &= e_1(nT) + S_2(e_2(nT)), \\ u_2(nT) &= -u(nT-T) = u(nT) - K_u \Delta u(nT) \\ &= e_2(nT) + S_1(e_1(nT)). \end{cases} \qquad (6.87)$$

Observe, moreover, that when $e(nT)$ and $r(nT)$ are in the regions IC1, IC2, IC5, or IC6, one has

$$\| S_1(e_1(nT)) \| \; = \; \left\| - \frac{K_u L}{2(2L - K_i \, | \, e(nT) \, |)} \times \right.$$

$$\left. \left[\left(K_i + \frac{K_p}{T} \right) e_1(nT) \; - \; \frac{K_p}{T} e_1(nT - T) \right] \right\|$$

$$\leq \frac{K_u L}{2(2L - K_i M_e)} \left[\left(K_i + \frac{K_p}{T} \right) | \, e_1(nT) \, | \; - \; \frac{K_p}{T} M_e \right]$$

$$= \frac{K_u K_p M_e L}{2(2L - K_i M_e)} \; + \; \frac{K_u K_i + K_p L}{2(2L - K_i M_e)} \, | e(nT) |,$$

$$(6.88)$$

and

$$\| S_2(e_2(nT)) \| \leq \| N \| \cdot | e_2(nT) |, \qquad (6.89)$$

where $\| N \|$ is the operator norm of the given $N(\cdot)$, or the gain of the given nonlinear system, defined as in formula (6.78), and M_e is the maximum magnitude of the error signal

$$M_e := \max \left\{ \sup_{n \geq 0} | e(nT) |, \; \sup_{n \geq 1} | e(nT - T) | \right\}. \qquad (6.90)$$

In regions IC1-IC8, $K_i M_e \leq L$.

To this end, an application of the Small Gain Theorem (Theorem 6.1) produces the following sufficient condition for the BIBO stability of the closed-loop nonlinear fuzzy PI control system:

$$\frac{K_u (T K_i + K_p)}{2T} \| N \| < 1. \qquad (6.91)$$

When $e(nT)$ and $r(nT)$ are in the regions IC3, IC4, IC7, or IC8, one similarly obtains a sufficient stability condition as follows:

$$\frac{K_u (T K_i + K_p)}{2T} \| N \| < 1. \qquad (6.92)$$

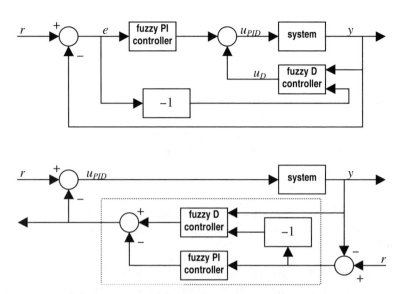

Figure 6.44 Equivalent closed-loop control systems

When $e(nT)$ and $r(nT)$ are in the rest of the regions, from IC9 to IC20, the other control laws are used. In this case, the stability conditions are found to be

$$
\begin{cases}
\dfrac{K_u K_i L}{2T} \| N \| < 1, \\[2ex]
\dfrac{K_u K_p L}{2T} \| N \| < 1, \\[2ex]
\| N \| \text{ is bounded.}
\end{cases}
\tag{6.93}
$$

By combining all the above conditions together, one arrives at the following result.

Theorem 6.4. A sufficient condition for the nonlinear fuzzy PI control system shown in Figure 6.23 to be globally BIBO stable is that (i) the given nonlinear system has a bounded norm (gain)

$\| N \| < \infty$, and (ii) the parameters of the fuzzy PI controller, K_p, K_d, and K_u, satisfy

$$\frac{K_u(\gamma K_i + K_p)}{2T} \| N \| < 1, \tag{6.94}$$

where $\gamma = \max\{L, T\}$.

D. BIBO Stability of Fuzzy PI+D Control Systems

The stability analysis of fuzzy PI+D control systems is in a sense to combine the results obtained individually for fuzzy PI and PD control systems.

Consider the fuzzy PI+D control system shown in Figure 6.27. First, observe that if one disconnects the fuzzy D controller from Figure 6.27, one has the fuzzy PI control system, exactly the same as Figure 6.23. Hence, all the results obtained in the last subsection for the fuzzy PI control system apply to this situation.

For the fuzzy PI+D control system shown in Figure 6.27, one can easily verify that it is equivalent to either one of the two configurations shown in Figure 6.44. Recall from the Small Gain Theorem (Theorem 6.1) that if one let the system denoted by S_1 and the fuzzy PI+D controller together be denoted by S_2, which is the dashed box in the second picture of Figure 6.44, then one can obtain a sufficient condition for the BIBO stability of the overall closed-loop control system from the bounds

$$\begin{cases} \| S_1(u_{\text{PID}}) \| \le M_1 + L_1 \| u_{\text{PID}} \|, \\ \| S_2\left(\begin{bmatrix} y \\ e \end{bmatrix}\right) \| \le M_2 + L_2 \left\| \begin{bmatrix} y \\ e \end{bmatrix} \right\|, \end{cases} \tag{6.95}$$

where M_1, M_2, L_1, L_2 are constants, with $L_1 L_2 < 1$, as discussed in detail in Theorem 6.1.

Here, observe that due to the special structure of the D controller, denoted S_D, one actually has

$$\left\| S_2\left(\begin{bmatrix} y \\ 3 \end{bmatrix}\right) \right\| = \left\| \begin{bmatrix} S_D & -1 \\ 0 & S_{\text{PI}} \end{bmatrix} \begin{bmatrix} y \\ e \end{bmatrix} \right\|$$

$$\le \left\| \begin{bmatrix} S_D & -1 \\ 0 & S_{\text{PI}} \end{bmatrix} \right\| \cdot \left\| \begin{bmatrix} y \\ e \end{bmatrix} \right\|$$

$$\le \max\{ \| S_D \|, \| S_{\text{PI}} \|, 1 \} \cdot \max\{ \| y \|, \| e \| \}.$$

Hence, a sufficient condition for the overall fuzzy PI+D control system to be BIBO stable is the worst one between the fuzzy PI and D control systems. Namely, one may use the larger norms from the right-hand side of the above. Thus, one will have the second equation of inequalities (6.95) in which S_D and S_{PI} are separated in M_2 and/or L_2, so that the analysis performed in the last two subsections can be repeated here for inequalities (6.95).

Note that in this case one may assume that max{ $\|S_D\|$, $\|S_{PI}\|$ } ≥ 1; otherwise, the system will be stable without additional conditions by the contraction mapping principle. Under this inequality, the second condition of (6.95) can be guaranteed.

E. Graphical Stability Analysis of Fuzzy PID Control Systems

It has been shown how to analyze the BIBO stability of a fuzzy PID type of control system in the last few sections. Sufficient conditions so obtained are useful for controllers design. Although such sufficient conditions are generally conservative, just like many other sufficient conditions derived via different methods for nonlinear systems, they provide useful guidelines for designing "safe" controllers (namely, stabilizing controllers with desirable robustness against system parameter variation and/or external disturbances).

In this section, a graphical approach is introduced to the BIBO stability analysis for PID type of fuzzy control systems. Only the fuzzy PI+D control systems shown in Figure 6.27 will be discussed, but the methodology clearly is applicable to different types of fuzzy PID control systems.

In Figure 6.27, observe that the control signal $u_{PID}(nT)$ to the error signal $e(nT)$ can be related implicitly through the closed-loop configuration. To find the relation in the z-domain, let

$$f(z) = \frac{e(z)}{u_{PID}(z)} \tag{6.96}$$

and

$$g(z) = \frac{v(z)}{u_{PID}(z)}, \tag{6.97}$$

where

$$f(z) = f_0 + f_1 z^{-1} + f_2 z^{-2} + \dots$$

$$g(z) = g_0 + g_1 z^{-1} + g_2 z^{-2} + \dots$$

are unknown (not explicitly known) but well-defined, and similarly,

$$e(z) = e_0 + e_1 z^{-1} + e_2 z^{-2} + \dots$$

$$v(z) = v_0 + v_1 z^{-1} + v_2 z^{-2} + \dots$$

$$u_{\text{PID}}(z) = u_0 + u_1 z^{-1} + u_2 z^{-2} + \dots .$$

Let

$$\| F \| = \sum_{i=1}^{\infty} f_i , \qquad \| G \| = \sum_{i=1}^{\infty} g_i ,$$

$$\| E \| = \sum_{i=1}^{\infty} e_i , \qquad \| V \| = \sum_{i=1}^{\infty} v_i , \qquad \| U \| = \sum_{i=1}^{\infty} u_i .$$

For the BIBO stability of the closed-loop system, it must require that both $\| F \|$ and $\| G \|$ be finite. So, let

$$\| F \| = \alpha_1,$$

$$\| G \| = \alpha_2,$$

where α_1 and α_2 are constants to be determined. Then, one has

$$\| E \| \le \alpha_1 \| U \|, \tag{6.98}$$

$$\| V \| \le \alpha_2 \| U \|. \tag{6.99}$$

Suppose that

$$y(nT+T) = N(y(nT)) + H_1(u_{\text{PID}}(nT)) + H_2(w(nT)), \tag{6.100}$$

where $\{w(nT)\}$ are disturbances ($H_2 = 0$ if it does not exist), N is the closed-loop system operator, and N, H_1, H_2 are bounded nonlinear functions (in operator norm):

$$\| N \| < \infty, \qquad \| H_1 \| < \infty, \qquad \| H_2 \| < \infty.$$

Then, it follows that

$$\| H_1(u_{\text{PID}}) \| = \| y(nT+T) - N (y(nT)) - H_2(w(nT)) \|$$

$$\leq \ \| y(nT+T) \| + \| N \| \cdot \| y(nT) \| + \| H_2 \| \cdot \| w \|.$$

Since $\| H_1(u_{\mathrm{PID}}) \| \leq \| H_1 \| \cdot \| u_{\mathrm{PID}} \|$, to ensure the left-hand side be bounded, one can require

$$\| H_1 \| \cdot \| u_{\mathrm{PID}} \| \leq \| y(nT+T) \| + \| N \| \cdot \| y(nT) \| + \| H_2 \| \cdot \| w \|$$

or, even more conservatively,

$$\| u_{\mathrm{PID}} \| \leq \beta_1 + \beta_2 \| E \| + \beta_3 \| V \|, \tag{6.101}$$

where

$$\beta_1 \leq \frac{1}{\| H_1 \|} \, (\| y(nT+T) \| + \| N \| \cdot \| y(nT) \| + \| H_2 \| \cdot \| w \|),$$

$$\beta_2 \leq \max\{ \| K_{\mathrm{P}} \|, 1 \},$$

$$\beta_3 \leq \max\{ \| K_{\mathrm{I}} \|, \| K_{\mathrm{D}} \| \}.$$

Substituting (6.101) into (6.98) gives

$$\| E \| \leq \alpha_1 (\beta_1 + \beta_2 \| E \| + \beta_3 \| V \|)$$

which yields (if $\alpha_1 \beta_2 < 1$)

$$\| V \| \geq \frac{1}{\alpha_1 \beta_3} [(1 - \alpha_1 \beta_2) \| E \| - \alpha_1 \beta_1]. \tag{6.102}$$

Similarly, substituting (6.101) into (6.99) leads to (if $\alpha_2 \beta_3 < 1$)

$$\| V \| \leq \frac{1}{1 - \alpha_2 \beta_3} [\alpha_2 \beta_2 \| E \| + \alpha_2 \beta_1]. \tag{6.103}$$

It is clear that the two straight lines on the right-hand side of (6.102) and (6.103) will create a common region for $\| V \|$ in the $\| E \|$ - $\| V \|$ plane. If $\alpha_1 \beta_2 > 1$ (the first straight line has a negative slope) and $\alpha_2 \beta_3 < 1$ (the second straight line has a positive slope), then an open region is created as shown in Figure 6.45. In this case, there is no common region in which the errors e and v are guaranteed to be bounded. Therefore, this case should be avoided.

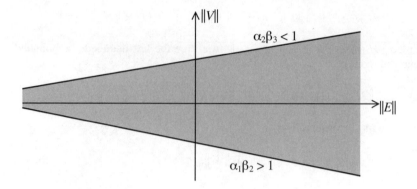

Figure 6.45 An open common region in the error space

· Observe that one is dealing with norms, so only the first quadrant of the $\| E \|$ - $\| V \|$ plane is intersecting. To obtain a closed region on the plane, one can equate (6.102) and (6.103), which gives

$$\frac{1}{\alpha_1\beta_3}[\,(1-\alpha_1\beta_2)\,\| E_r \| - \alpha_1\beta_1\,] = \frac{1}{1-\alpha_2\beta_3}[\,\alpha_2\beta_2\,\| E_r \| + \alpha_2\beta_1\,],$$

(6.104)

where $\| E_r \|$ is the value of $\| E \|$ at a point denoted by r on the plane. Solving (6.104) for $\| E_r \|$, one obtains

$$\| E_r \| = \frac{\alpha_1\beta_1}{1-(\alpha_1\beta_2 + \alpha_2\beta_3)}.$$

(6.105)

Substituting (6.105) into (6.103) yields the corresponding

$$\| V_r \| = \frac{\alpha_2\beta_1}{1-(\alpha_1\beta_2 + \alpha_2\beta_3)}.$$

(6.106)

Therefore, the intersection point r of the two straight lines has the coordinates

$$(\,\| E_r \|\,,\, \| V_r \|\,) = \left(\frac{\alpha_1\beta_1}{1-(\alpha_1\beta_2 + \alpha_2\beta_3)}, \frac{\alpha_2\beta_1}{1-(\alpha_1\beta_2 + \alpha_2\beta_3)} \right),$$

(6.107)

which are finite if $\alpha_1\beta_2 + \alpha_2\beta_3 < 1$. This situation is shown in Figure 6.46.

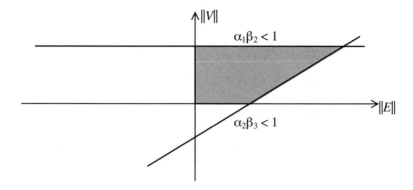

Figure 6.46 A closed common region in the error space

In summary, one has the following sufficient conditions.

Theorem 6.5. A sufficient condition for the BIBO stability of the fuzzy PI+D control system is that the fuzzy control gains K_P, K_I, K_D are chosen such that there are constants α_1, α_2, β_1, β_2, β_3 that together satisfy

(i) $\| E \| \leq \alpha_1 \| U \|, \quad \| V \| \leq \alpha_2 \| U \|,$

(ii) $\beta_1 \leq \dfrac{1}{\| H_1 \|} \ [\ (1 + \| N \|) \cdot \| y \| + \| H_2 \| \cdot \| w \| \ ],$

(iii) $\beta_2 \leq \max\{ \ \| K_P \|, 1 \ \},$

(iv) $\beta_3 \leq \max\{ \ \| K_I \|, \| K_D \| \ \},$

(v) $\alpha_1 \beta_2 < 1, \quad \alpha_2 \beta_3 < 1,$

(vi) $\alpha_1 \beta_2 + \alpha_2 \beta_3 < 1,$

and the closed region shown in Figure 6.46 is not empty. Here,

$$\| y \| = \sup_{n \geq 0} \ \| y(nT) \|.$$

Finally, it is remarked that the norm $\| y \|$ in condition (ii) of Theorem 6.5 needs not be finite when verifying the constant β_1; namely, if $\| y \| = \infty$ then condition (ii) is trivially satisfied. However, if all the other conditions are satisfied simultaneously, then $\| y \|$ would be finite as a result of the BIBO stability of the overall control system.

Problems

P6.1   Consider the PD-control system shown in Figure 6.47, where the given linear system has a first-order transfer function, $1/(as + b)$, with known constants $a > 0$ and b. Design the PD controller by means of determining the two constant control gains, K_P and K_D, such that the system output can track the set-point while the entire feedback control system is stable.

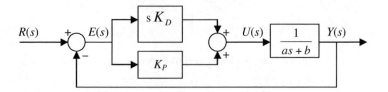

Figure 6.47 A PD control system

P6.2   Consider the PI-control system shown in Figure 6.48, where the given linear system has a first-order transfer function, $1/(as^2 + b)$, with known constants $a > 0$ and b. Design the PI controller by means of determining the two constant control gains, K_P and K_I, such that the system output can track the set-point while the entire feedback control system is stable.

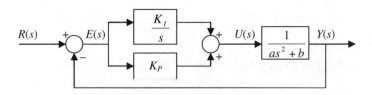

Figure 6.48 A PI control system

P6.3   Consider the analog D-controller

$$u(t) = K_D \frac{d}{dt} e(t)$$

which, in the frequency domain, is

$$U(s) = sK_D E(s)$$

Use the bilinear transform

$$s = \frac{2}{T} \frac{z-1}{z+1}$$

to convert this analog D-controller to be a digital one. Show detailed derivation and final result of its digital formula and draw the corresponding digital block diagram.

P6.4   Repeat all numerical simulations shown in Example 6.5, so as to experience how and how well the fuzzy PD controller works.

P6.5   Perform some numerical simulations for the fuzzy PI controller and compare them with the conventional PI controller. You may use the same models as those simulated in Example 6.6.

P6.6   Perform the graphical stability analysis studied in Section III-E on the fuzzy PD and fuzzy PI control systems individually.

CHAPTER 7

Computational Verb Fuzzy Controllers

As seen from the previous chapters, the traditional fuzzy logic applications have been developed around a fuzzy rule of the form "IF (condition) - THEN (conclusion)." In this framework, both the condition and conclusion normally consist of a logical operation over a set of simple logical statements, with each statement in the form of "variable x belongs to set $S_{\text{description}}$." This form of logical statement can be intuitively translated in human language as "variable x is (some description)," where the main verb is "*is*," and the description is some vague quantitative representation, e.g., small, medium, large, etc.

There are many applications where fuzzy logic is found more useful if each statement in the fuzzy rule can be extended from the form "variable x is (some description)" to containing a more descriptive verb other than the very special *static* verb "*is*." For example, a statement "variable x *increases* to (some description)," where the *dynamic* verb "increases," can easily extend the application scope of the fuzzy rule to a much wider range. Given that a statement with a particular verb such as "*is*" can be converted to a more complicated but standard statement with a general verb such as "*increases*" via some additional mathematical operations, advanced fuzzy logic allows us to better understand and use fuzzy rules with common knowledge of their spoken languages.

This chapter studies an extension of the traditional fuzzy rules converting the static verb "*is*" into a more general scenario where a dynamic verb can be used to better describe the underlying information processing. This extension is first presented with mathematical analysis, showing certain equivalence between the dynamic verb "*becomes*" and the static verb "*is*" and an extension of the former from the latter. The concept of organizing a group of rules, each containing the verb "is," into a single rule containing a more general dynamic verb, such as "becomes," is then introduced to reduce the corresponding rule base. This concept is furthermore applied to the fuzzy PID controllers for linear systems. Finally, a design algorithm and some numerical simulations are presented to illustrate this new concept and its applications.

I. COMPUTATIONAL VERBS AND VERB NUMBERS

A typical fuzzy control rule has the IF-THEN form that intuitively contains the special static verb "is" to describe a variable. For example, consider the rule

$$\text{IF } x \in S_{\text{high}} \text{ THEN } y \in S_{\text{small}},$$

where x is a variable representing some physical property such as temperature, S_{high} is the fuzzy set describing some high values, y is the control variable to a particularly given system, and S_{small} is the fuzzy set describing some low values. This rule is very typical in fuzzy control applications, as seen from the previous chapters. This rule can be interpreted by investigators in the following equivalent linguistic form:

"IF the temperature *is* high THEN the control action *is* small."

If one wants to generalize the verbs "*is*" in this rule into any other verb, the problem of modeling these more active types of verbs must be carefully addressed. For example, one may try to extend the above fuzzy rule to

"IF the temperature *increases* (to) a high value THEN the control action *decreases* (by) a small value."

In doing so, however, it is clear that the meaning (and the consequent action) of this fuzzy rule has been changed significantly, with the change of the verb from "*is*" to "*increases*" and "*decreases*," which changes the static description "being" (the status) to a dynamic description "becoming" (a process).

A systematic theoretical framework for modeling verbs is called *computational verb theory*, which is first reviewed below.

A. Fundamental Concepts

From the linguistic point of view, two essential grammatical components of a natural language are nouns and verbs. Nouns are used to model (or represent) a static state associated with "*being*" while verbs are used to model (or describe) a transitioning state associated with "*becoming*." Notably, the special verb "be" is also used to represent "being" which, however, differs from a noun in that a noun represents an entity itself while the verb "be" represents its status.

To solve engineering problems, fuzzy theory and probability are used to implement adverbs-adjectives and noun-phrases into computers. Similarly, computational verbs are used to implement verbs into computers. In fuzzy logic theory, membership functions are used to numerically model (adverbs and) adjectives such as "(very) large," "(somewhat) small," and so on. In contrast to this traditional approach, evolving functions are used to numerically model verbs in the computational verb theory.

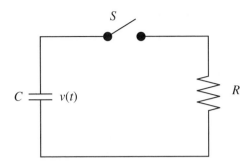

Figure 7.1. Block diagram of a simple RC circuit
with $R > 0$ and $C > 0$

The scientific definition of a general class of computational verbs is too complex to be operational in engineering applications. Therefore, only some very specific cases are discussed here in this section. In the following, an explicit mathematical definition of some specific computational verbs is presented and then used throughout the chapter.

Definition 7.1. A computational verb V is defined by the following evolving function:

$$\varepsilon_V : T \times \Omega \to \Omega, \tag{7.1}$$

where $T \in R$ and $\Omega \in R^n$ are the time and the universe of discourse, respectively.

The following example illustrates the above definition of computational verbs within the context of an engineering problem.

Example 7.1.

Consider the simple RC-circuit shown in Figure 7.1. When the switch S is on, the capacitor C will be discharged such that the voltage $v(t)$ will eventually approach zero. Assuming that the switch S is being turned on initially, at $t_0 = 0$, and the initial voltage across the capacitor is $v(t_0) = v_0 > 0$. Then, the following sentence can be used to describe the changing process of the voltage across R: The voltage $v(t)$ decreases from v_0 to 0. According to the physical law, the evolving function for the verb "decrease" in this sentence is given by

$$\varepsilon_{\text{decrease}}(t,v) = v_0 \exp\left\{-\frac{1}{RC}t\right\}. \tag{7.2}$$

Table 7.1 Examples of Different Types of Computational Verb Numbers

Verb numbers	Types
become 3.12	real verb number
belong to $[-1, 1]$	interval verb number
approach $3+ i4$, $i^2 =1$	complex verb number
drop to very low	fuzzy verb number
stay at 3.12	real singleton verb number
remain inside $[-1, 1]$	interval singleton verb number
be very low	fuzzy singleton verb number
believe in truth	Boolean verb number

Table 7.2 Examples of Verb Statements and Canonical Forms in *become*

Verb statements	Canonical forms in *become* $(\cdot,\cdot)$
$\|e(k)\|$ decreases	become (current, less than current)
$\|e(k)\|$ decreases to small	become (current, small)
$\|e(k)\|$ increases to ten	become (current less than ten, ten)
$\|e(k)\|$ increases from small to big	become (small, big)
$\|e(k)\|$ decreases from 10 to 0.01	become (10, 0.01)
$\|e(k)\|$ climbs very fast	(very fast) ∘ become (current, bigger than current)
$\|e(k)\|$ jumps up	(very fast) ∘ become (current, bigger than current)
$\|e(k)\|$ bounces back	become (current, before current)
$\|e(k)\|$ stays high	become (high, high)
$\|e(k)\|$ is high	become (high, high)

It is clear that a typical fuzzy rule based on the static verb "is" is not easy, if ever possible, to describe this dynamical process of the voltage change.

B. Computational Verb Numbers

Since a dynamic verb must evolve within some state space, which can be either physical or mental, the characteristics of the state space determine the nature of the computational verbs. When the state space is numerical, the corresponding computational verb is called a *computational verb number* (*verb number,* for short). Some examples of verb numbers are listed in Table 7.1. A fuzzy number is a special kind of verb number, a static verb number, called *fuzzy singleton verb number.*

In order to minimize the number of different verbs used in a verb-based controller, one needs to choose a few canonical computational verbs. From the (fuzzy) PID controllers design examples discussed in Chapter 6, it is easy to imagine that in verb-based PID controllers, only one canonical computational verb, namely, *become*, is needed. The canonical form of the verb *become* is denoted as a function "*become*(state 1, state 2)," in which state 1 and state 2 can be crisp or fuzzy numbers or other special values, representing the process of change from state 1 to state 2 of the variable in interest. Some examples of the verb *become* are listed in Table 7.2, where the symbol "∘" denotes an operation between an adverb and a computational verb, in which the adverb can be quantified by a fuzzy membership value (as the weight) therefore the symbol "∘" can be just the numerical multiplication in this case.

Also observe that, as a special case of verb statements, the fuzzy statements in the last two rows of Table 7.2 can indeed be represented by canonical forms in *become*.

C. Verb Similarity

The similarity between verbs (verb similarity, for short) is of essential importance to the inference of verb rules. Since there is no crisp definition of similarity between two dynamic processes, the verb similarity may be defined based on different concerns. Here, rather than giving a precise definition for verb similarity, some criteria are given instead, as follows.

Definition 7.2. Given two computational verbs, V_1 and V_2, the verb similarity $S(V_1,V_2)$ should satisfy the following properties:

1. $S(V_1,V_2) \in [0,1]$;

2. $S(V_1,V_2) = S(V_2,V_1)$;

3. $S(V_1,V_2) = 1$ if and only if $V_1 = V_2$, which means that both computational verbs have the same evolving function.

For the purpose of designing digital computational verb-based controllers, consider the cases where the evolving functions of the computational verbs are discrete-time samples.

Let two computational verbs V_1 and V_2 be associated, respectively, with normalized evolving functions $\varepsilon_{V_1}(k) \in [0,1]$, $\varepsilon_{V_2}(k) \in [0,1]$, $\forall k \in Z$. Then, the most commonly used verb similarity is computed by

$$S(V_1,V_2) = \begin{cases} \dfrac{\displaystyle\sum_{k=0}^{n} \varepsilon_{V_1}(k) \wedge \varepsilon_{V_2}(k)}{\displaystyle\sum_{k=0}^{n} \varepsilon_{V_1}(k) \vee \varepsilon_{V_2}(k)}, & \displaystyle\sum_{k=0}^{n} \varepsilon_{V_1}(k) \vee \varepsilon_{V_2}(k) \neq 0, \\[2em] 0, & \text{otherwise.} \end{cases}$$ (7.3)

II. VERB RULES AND VERB INFERENCE

A. Verb Inference with a Single Verb Rule

The process of evaluating a verb statement is called *verb inference*. In the design of computational verb-based controllers, one uses *verb generalized modus ponens* (GMP). In verb GMP, when a verb IF-THEN rule and its antecedent, which is a verb statement, are approximately matched, a consequent, which is also a verb statement, may be inferred.

More precisely, if only one verb rule is considered, then a verb GMP can be formally written as

IF NP_1 V_{x0}, THEN NP_2 V_{y0}.

$\underline{NP_1\ V_1}$

$NP_2\ V_2$ (7.4)

where NP_1 and NP_2 denote two noun-phrases and V_1 is an observed verb matching the antecedent computational verb to a verb similarity. All the computational verbs located above the horizontal line are considered given and known, whereas the computational verbs blow the line are considered unknown.

A (computational) verb rule can be analytically represented by a verb implication relation. In verb relations, one considers *pairs* or more generally *n-tuples* related to a *degree of verb similarity*. In addition to the question of whether some verb belonging to a verb set can be considered as a matter of degree of verb similarity; whether some verbs are associated with each other may also be considered as a matter of degree of verb similarity. In computational verb theory, computational verb relations play a role in a way similar to what the relation functions do in conventional approaches.

A computational verb relation R is defined over the Cartesian product of two or more verb sets. For the time being consider only verb relations defined over the Cartesian product of two verb sets, V_x and V_y. Assume that both verb sets have only finite numbers of computational verbs, denoted as

$$V_x = \{ V_{x0}, V_{x1}, \ldots, V_{xm} \}, \tag{7.5a}$$

$$V_y = \{ V_{y0}, V_{y1}, \ldots, V_{yn} \}. \tag{7.5b}$$

Let a binary verb relation R be defined on $V_x \times V_y$. Then, one can list all its pairs explicitly as follows:

$$R: V_x \circ V_y \;=\; \{ ((V_x,V_y), \sigma_R(S_x(V_{x0},V_x),S_y(V_{y0},V_y))) \}$$

$$= \sum_{(V_x,V_y)\in V_x\times V_y} \sigma_R(S_x(V_{x0},V_x),S_y(V_{y0},V_y)) \,/\, (V_x,V_y), \tag{7.6}$$

where every pair (V_x,V_y) belongs to $V_x \times V_y$, $S_x(\cdot,\cdot)$ and $S_y(\cdot,\cdot)$ are two verb similarities, V_{x0} and V_{y0} are *reference computational verbs*, and $\sigma_R: [0,1] \times [0,1] \rightarrow [0,1]$ is called a *verb relation*. The most commonly used verb relation is

$$\sigma_R(x,y) = x \wedge y. \tag{7.7}$$

In order to perform verb GMP in relation (7.4), over $V_1 \circ V_2$, the following mathematical formulas behind the verb composition are used:

$$S_y(V_2,V_y) = \bigoplus_{V_x \in V_x} S_x(V_1,V_x) \otimes \sigma_R(\, S_x(V_{x0},V_x), S_y(V_{y0},V_y) \,),$$

$$\forall V_y \in V_y, \tag{7.8}$$

where $\oplus$ and $\otimes$ denote the so-called *s*-norm and *t*-norm, respectively, usually defined by $\oplus := \max$, $\otimes := \min$. The result of (7.8) is the following *similarity function*:

$$V_2 = \sum_{V_y \in V_y} S_y(V_2,V_y)/V_y. \tag{7.9}$$

Observe that the unknown verb V_2 is implicitly represented by the verb similarities between V_2 and all verbs in the verb set V_y. If one chooses $\oplus = \max$ and $\otimes = \min$, as mentioned above, and $\sigma_R(\cdot,\cdot)$ the same as in (7.7) with $\wedge = \min$ therein, then the steps to implement verb GMP are listed as follows.

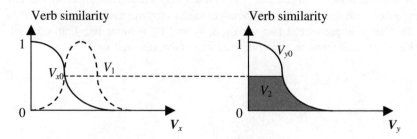

Figure 7.2. Illustration of the procedure of verb inference using verb GMP

Step 1.   For each $V_x \in V_x$, find the verb similarity $S_x(V_1,V_x)$ and represent V_1 by using the following similarity function:

$$V_1 = \sum_{V_x \in V_x} S_x(V_1,V_x)/V_x .$$ (7.10)

Step 2.   Use (7.8) to compute

$$S_y(V_2,V_y) = \max_{V_x \in V_x} \min(S_x(V_1,V_x), S_x(V_{x0},V_x), S_y(V_{y0},V_y))$$

$$= \min(S_y(V_{y0},V_y), \max_{V_x \in V_x} \min(S_x(V_1,V_x), S_x(V_{x0},V_x))),$$

$\forall V_y \in V_y.$

Obtaining

$$\alpha = \max_{V_x \in V_x} \min(S_x(V_1,V_x), S_x(V_{x0},V_x)),$$

as shown in the left part of Figure 7.2. This quantity is called the *firing level* of this verb rule.

Step 3.   Compute V_2 by

$$S_y(V_2,V_y) = \min(S_x(V_1,V_x), \alpha), \quad \forall V_y \in V_y,$$

as shown in the right part of Figure 7.2, where a function is considered as an adverb.

Step 4. The evolving function of V_2 can be constructed using a *deverbification* operation.

Figure 7.2 illustrates the procedure of verb inference using verb GMP. The horizontal axis representing the verb set V_x or V_y. The vertical axis denotes the degree of verb similarity. The shadowed region in the right-hand side of Figure 7.2 represents the similarity function of V_2.

B. Verb Inference with a Verb Algorithm

A *verb algorithm* consists of a set of verb IF-THEN rules, which are defined over the same product space and are connected by the connective ELSE. Corresponding to different verb relation operators, R, the ELSE can be interpreted as AND if $R(0) = 1$, OR if $R(0) = 0$.

To generate control signals, one usually considers the following GMP based on a verb algorithm:

$$\text{IF NP}_1 \ V_x^{(1)} \ \text{THEN NP}_2 \ V_y^{(1)} \ \text{ELSE}$$

$$\text{IF NP}_1 \ V_x^{(2)} \ \text{THEN NP}_2 \ V_y^{(2)} \ \text{ELSE}$$

$$\vdots$$

$$\text{IF NP}_1 \ V_x^{(l)} \ \text{THEN NP}_2 \ V_y^{(l)} \ \text{ELSE}$$

$$\underline{\text{NP}_1 \ V_1}$$

$$\text{NP}_2 \ V_2 \tag{7.11}$$

If one chooses the verb relation operator (7.7), then

$$V_2 = \bigcup_{i=1}^{l} V_1 \circ R_i \ ,$$

where R_i is the verb implication relation of the i-th verb rule.

One example for a verb algorithm with 3 rules is shown in Figure 7.3.

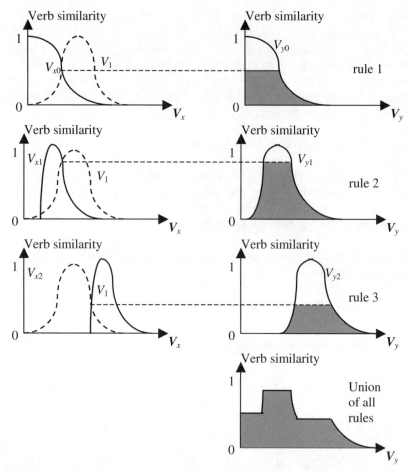

Figure 7.3. Illustration of the procedure of verb GMP based on a 3-rule verb algorithm

Observe that the final result of the verb GMP is a verb similarity function. In many applications, one needs to reconstruct dynamics from verb similarity functions, as further discussed below.

C. Deverbification: Reconstruct Computational Verbs from Similarity Functions

Although verb inference is based on verb similarity and is only one way to perform verb inference, this method is very efficient when the underlying dynamics change their qualitative behaviors continuously with respect to parameter variations. Since in this method all verbs are represented by their verb similarity functions, the resulting verbs are also represented by verb similarity functions. However, in many applications, such as computational verb-based controllers, verb similarity functions cannot be directly used to generate control signals. Instead, a verb similarity function has to be translated back to a computational verb. This process is called *deverbification*.

Assume that there are K verb rules in the verb algorithm. Then, V_2 is given by

$$V_2 = \sum_{p=1}^{K} \alpha_p \circ V_y^p \, , \qquad (7.12)$$

where α_p is the firing level (i.e., the weight) of the p-th verb rule. This is a typical deverbification formula.

III. COMPUTATIONAL VERB-BASED FUZZY PID CONTROLLERS

Recall the notion of the fuzzy *proportional-integral-derivative* (PID) controllers studied in Chapter 6. The block diagram of a typical fuzzy PID controller is shown in Figure 7.4, where $r(t)$ is the *reference signal,* $y(t)$ is the *output,* $e(t) = r(t) - y(t)$ is the *control error,* and $u(t)$ is the *control signal.*

As discussed in detail in Chapter 6, a typical conventional or type-1 fuzzy PID-controller consists of a P-controller with a parameter $K_p(t)$ called *proportional gain,* an I-controller with a parameter $K_i(t)$ called *integral gain,* and a D-controller with a parameter $K_d(t)$ called *derivative gain.* The conventional PID control law is defined by a mapping between $u(t)$ and $e(t)$ and is given by the following linear combination in the time domain:

$$u(t) = K_p(t)e(t) + K_i(t) \int_0^t e(\tau)d\tau \ + K_d(t)\frac{d}{dt}e(t) \, . \qquad (7.13)$$

If one implements the PID controllers by using digital processors, then other than the digital PID controllers studied in Chapter 6, the following time-varying-gain discrete-time PID controller is also very useful:

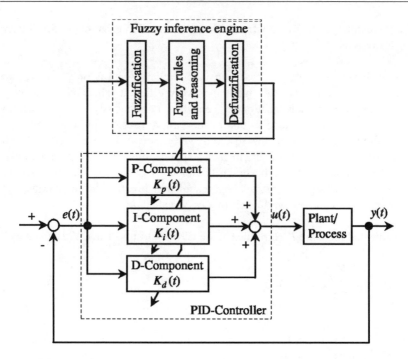

Figure 7.4. The block diagram of a type-1 fuzzy PID controller

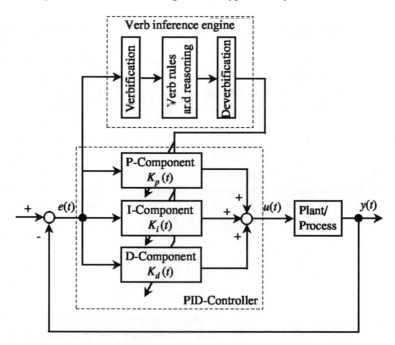

Figure 7.5. Block diagram of computational verb PID controller

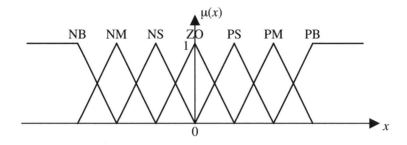

Figure 7.6. The membership functions of fuzzy numbers modeling A_i and B_i

$$u(k) = K_p(k)e(k) + K_i(k)T_s \sum_{i=0}^{k} e(i) + \frac{K_d(k)}{T_s} \Delta e(k). \tag{7.14}$$

where T_s is the sampling period, $u(k)$ and $e(k)$ are the control signal and the error at time kT_s, respectively, $K_p(k)$, $K_i(k)$, and $K_d(k)$ are the proportional, integral and derivative gains at time kT_s, respectively. As usual, define the *change of error* during the sampling interval $[(k-1)T_s, kT_s]$ as $\Delta e(k) = e(k) - e(k-1)$.

Observe in Figure 7.4 that the parameters of the type-1 fuzzy PID controller are tuned by the fuzzy inference engine, as discussed in Chapter 6.

The computational verb-based fuzzy PID controller can be constructed as shown in Figure 7.5. The similarity between a fuzzy PID controller and a verb PID controller results from the fact that the only verb used in fuzzy logic is "be" (i.e., *is*), a *static verb*. However, their design principles are very different; for example, the output of the "verbification" block is a verb – "*observe*".

A. Fuzzy Gain Schedulers

Without loss of generality, assume that the gains $K_p(k)$ and $K_d(k)$ can be tuned (scheduled) only when they are within intervals $[K_{p,\min}, K_{p,\max}]$ and $[K_{d,\min}, K_{d,\max}]$, respectively. While in a fuzzy gain scheduler (a fuzzy rule base), the PID gain parameters are determined by the current values of $e(k)$ and $\Delta e(k)$, due to the static nature of the special verb "*is*," in a verb gain scheduler the PID gain parameters are determined by both the historical values and the current value of $e(i)$, $i = 0,1,2,\ldots,k$, due to the dynamic nature of the general verb "*becomes*." The difference between fuzzy and verb gain

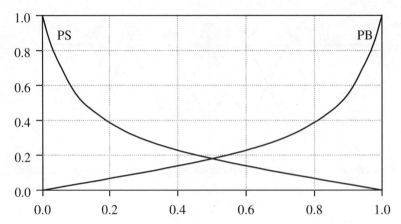

Figure 7.7. Membership functions of fuzzy sets
PS and PB for $\tilde{K}_p$ and $\tilde{K}_d$

schedulers paves the way to tune the gain parameters of the controller dynamically throughout the dynamical control process.

In order to avoid clutters, the gains K_p and K_d are normalized within the interval [0,1] via formulas

$$\tilde{K}_p = \frac{K_p - K_{p,\min}}{K_{p,\max} - K_{p,\min}}, \quad \tilde{K}_d = \frac{K_d - K_{d,\min}}{K_{d,\max} - K_{d,\min}}. \tag{7.15}$$

Furthermore, the value of K_i is defined indirectly by a parameter γ through the following relation:

$$K_i = \frac{K_p^2}{\gamma K_d}. \tag{7.16}$$

Since a verb-based PID controller can be readily generated from a type-1 fuzzy PID controller, it is helpful to briefly review the designing process of the fuzzy PID controller discussed in Chapter 6.

The fuzzy rule base, referred to as the gain scheduler here, tunes the three parameters $\tilde{K}_p$, $\tilde{K}_d$ and γ by using a set of m fuzzy rules, each is in the following form:

IF $e(k)$ is A_i AND $\Delta e(k)$ is B_i
THEN $\tilde{K}_p$ is C_i AND $\tilde{K}_d$ is D_i AND $\gamma = \gamma_i$, $i = 1, 2, ..., m$. (7.17)

where γ_i is a constant, A_i and B_i are fuzzy sets with membership functions given in Figure 7.6, and C_i and D_i are fuzzy sets with membership functions simply chosen to be wither "big" or "small" as shown in Figure 7.7 and given as follows:

$$\mu_{\text{big}}(x) \quad = \quad \min\left(1, -\frac{1}{4}\ln(1-x)\right), \tag{7.18a}$$

$$\mu_{\text{small}}(x) = \quad \min\left(1, -\frac{1}{4}\ln(x)\right), \quad x \in (0,1). \tag{7.18b}$$

All basic building blocks for the fuzzy PID controller are now in place, so one is ready to generalize the fuzzy control rules to the verb-based control rules.

B. From Fuzzy Control Rules to Verb-Based Control Rules

Consider the following (combined) fuzzy rules for turning PID controllers:

> IF $e(k)$ is PB AND $\Delta e(k)$ is ZO
> THEN $\tilde{K}_p$ is PB AND $\tilde{K}_d$ is PS AND $\gamma = 2$.   (7.19)

This (combined) fuzzy rule is established based on the knowledge and experience of a control engineer. In Figure 7.8, a typical system response is shown.

The control rule (7.19) is used to handle the situation in Region 1 shown in Figure 7.8, where a big control signal is needed in order to deduce $e(k)$ quickly. In this case, the PID controller needs to have a large K_p, a large K_i and a small K_d in order to produce a large control signal, where a smaller γ yields a big K_i.

To translate the fuzzy rule in (7.19) into a verb rule, it turns out to be easier to study the phase plot shown in Figure 7.9. As clearly shown in the phase plot and the step response in Figure 7.8, the plant only stays in Region 1 for a very short time interval before it enters Region 2.

The (combined) fuzzy control rule for Region 2 is given by

> IF $e(k)$ is PB AND $\Delta e(k)$ is NS
> THEN $\tilde{K}_p$ is PB AND $\tilde{K}_d$ is PS AND $\gamma = 2$.   (7.20)

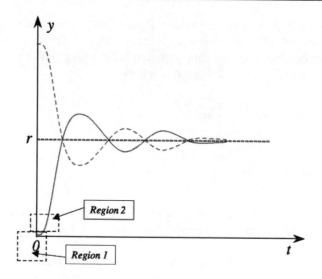

Figure 7.8. A typical step response of the system

Therefore, the actions in both fuzzy rules (7.19) and (7.20) are the same. It is now ready to use the following verb rule to implement the two fuzzy rules:

IF $e(k)$ *decreases* from PB

THEN $\tilde{K}_p$ is PB AND $\tilde{K}_d$ is PS AND $\gamma = 2$.   (7.21)

Here, one can see an advantage of using verb rules over fuzzy rules: two fuzzy rules are combined into one single verb rule. However, the main drawback of the verb rule (7.21) is that the conclusion of the rule is still not verbified. In order to verbify the conclusion statements, one needs to further construct some verb rules based on the global structure of the fuzzy rule space.

Observe that the fuzzy rule space of the fuzzy PID controller can be constructed as shown in Figure 7.10. This fuzzy rule space works only locally without considering the global behaviors of the trajectories of the control error $e(k)$. For example, the trajectories moving along clockwise directions constitute a global feature, which can be easily implemented by verb rules but cannot be reflected by the fuzzy rules. The aim, therefore, is to translate the fuzzy rule space shown in Figure 7.10 into three sets of verb rules.

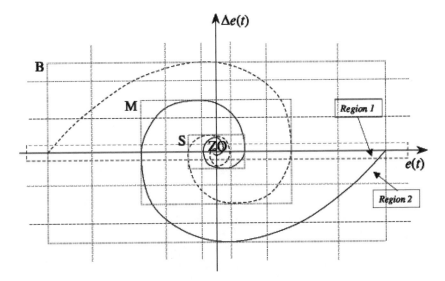

Figure 7.9. The phase plot of the step response of the system

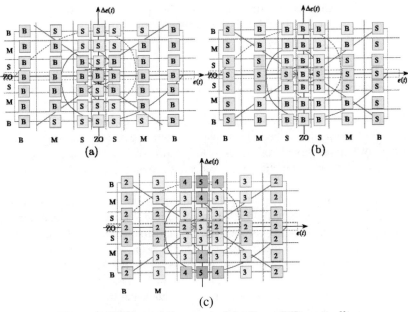

Figure 7.10. Fuzzy rule space of the fuzzy PID controller
presented together with phase space:
(a) Configuration of the rule set for $\tilde{K}_p$
(b) Configuration of the rule set for $\tilde{K}_d$
(c) Configuration of the rule set for γ

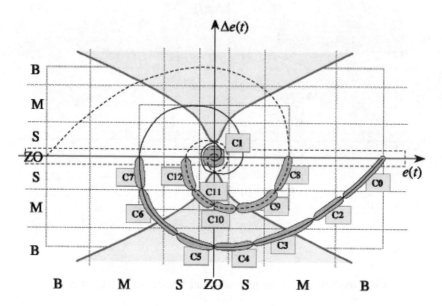

Figure 7.11. Chunks of trajectories in the phase space for designing computational verb P-controller

C. Constructing Verb Rules for Tuning the Gain $\tilde{K}_p$ from Phase Plots

When Figure 7.10 is examined by a human expert, especially when the two spiral trajectories are traced along, many dynamic patterns can be seen. The design of computational verb rules is to balance such details of the dynamic patterns and the global picture of the trajectories over the entire phase plot.

In order to clearly describe the analysis and design procedure, redraw Figure 7.10(a) in a continuous fashion as shown in Figure 7.11. Then, observe that in Figure 7.11 the boundaries between the big and small values of K_p have some qualitatively changes while many details have been removed so as to avoid possible confusion. A significant difference between Figures 7.10(a) and 7.11 is that, instead of chunking the phase plane, namely, the $(e, \Delta e)$-plane, by using fuzzy rules, one is now using verb rules to chunk the trajectories (the spiral curves). Since both fuzzy rules and verb rules describe only the *qualitative* behaviors of the plant, the accuracy of the boundaries between chunks is not important.

In Figure 7.11, all chunks along the trajectories are labeled by $C0, C1, \ldots$. For example, $C0$ corresponds to the following verb rule:

$C0$:   IF $|e(k)|$ *leaves* PB

      THEN $\tilde{K}_p$ *slowly decreases* from PB.   (7.22)

Since this is the very first verb rule generated from the phase plot based on static fuzzy logic knowledge, it is worthwhile to discuss the design principle about this rule in more detail.

Comparing the position of chunk $C0$ in Figure 7.11 with the fuzzy chunks in Figure 7.10(a), one can see that $C0$ overlaps with the governing regions of the following three fuzzy rules:

IF $e(k)$ is PB AND $\Delta e(k)$ is ZO THEN $\tilde{K}_p$ is PB,

IF $e(k)$ is PB AND $\Delta e(k)$ is NS THEN $\tilde{K}_p$ is PB,

IF $e(k)$ is PB AND $\Delta e(k)$ is NM THEN $\tilde{K}_p$ is PB.   (7.23)

It is easy to see that rules (7.23) are three *isolated* rules without connection, which actually exists as can be seen from the trajectories in the phase plot. The main reason is that this connection is generated by dynamics, which cannot be properly described by static noun phrases; namely, it cannot be represented by a standard fuzzy rule base. However, imagine that a trajectory is traveling along a spiral track on the phase plane; thus one can see that the trajectory is leaving the region where $|e(k)|$ is big and will arrive at the region where $|e(k)|$ is medium after a while. But it can also be imagined that the trajectory will not stay in the region where $|e(k)|$ is big for too long, although the entire chunk $C0$ is in such a region.

What will happen to $\tilde{K}_p$ when the trajectory is moving along the spiral curve in chunk $C0$? The rules in (7.23) only tell that $\tilde{K}_p$ will be big for all time. This is correct if one only looks at a small region around the trajectory. However, the global tendency in the phase plane shows that the spiral trajectory will eventually cross the boundary between the region where $\tilde{K}_p$ is PB and the region where $\tilde{K}_p$ is PS. Assuming that $\tilde{K}_p$ changes *continuously* on the phase plane, one will not accept the hypothesis that $\tilde{K}_p$ changes between PB and PS impulsively along a boundary. Instead, it would be more natural to accept the assumption that the changes are smooth from big to small along the chunks $C0$, $C2$ and $C3$. Hence, when the trajectory moves along chunk $C0$, $\tilde{K}_p$ will decrease continuously. The entire chunk $C0$ is in the big region and it will take the trajectory a relatively long time to move from the start to the end of chunk $C0$ because $\Delta e(k)$ is less than PM most of the time. Therefore, $\tilde{K}_p$ must decrease slowly.

Chunk $C2$ corresponds to the following verb rule:

$C2$:   IF $|e(k)|$ *decreases* from PB to PM

THEN $\tilde{K}_p$ *fast decreases* from PB.   (7.24)

The main difference between chunks $C0$ and $C2$ is that the entire $C0$ is in the region where $|e(k)|$ is PB while $C2$ crosses from the region where $|e(k)|$ is PB to the region where $|e(k)|$ is PM. Another difference is that, in chunk $C2$, $\Delta e(k)$ becomes bigger than that in chunk $C0$. Therefore, the decrease of $\tilde{K}_p$ is no longer *slow*.

Chunk $C3$ corresponds to the following verb rule:

$C3$:   IF $|e(k)|$ *decreases* from PM to PS

THEN $\tilde{K}_p$ *decreases* from PB to PS.   (7.25)

$C3$ is the first chunk to cross the boundary between the regions where $\tilde{K}_p$ is PB and is PS, respectively.

Chunks $C0$ to $C7$ cover the spiral trajectory in half of its period when it begins with a PB error. Chunks $C8$ to $C12$ cover the spiral trajectory in another half of its period when it begins with a PS error. Note that one can also define a series of such chunks for the case where the spiral trajectory begins with PVB (positive very big) error. Fortunately, in many cases one does not need to chunk the dynamics in such great detail for the following reasons:

1.   Chunks $C0$ to $C7$ already cover the governing ranges of all the detailed chunks. For example, chunks $C2$ and $C3$ cover chunks $C8$ and $C9$; chunk $C4$ covers $C10$; chunks $C5$ and $C6$ cover $C11$ and $C12$.

2.   The calculation of verb rules can tolerate an enormous amount of errors (noise and variations).

From the experience of designing a verb-based P-controller, it is known that only the rules corresponding to chunks $C0$ to $C7$ are sufficient for designing the gain scheduler for the P-controller. Furthermore, chunk $C0$ can be covered by chunk $C1$. Therefore, to design verb PID controllers, one only needs to implement rules that govern chunks $C1$ to $C7$. All verb rules corresponding to chunks $C1$ to $C7$ are listed in Table 7.3. The canonical forms in the *become* function of all rules shown in Table 7.3 are listed in Table 7.4.

Table 7.3. Verb Rules for Chunks $C1$ to $C7$ Shown in Fig. 7.11 for Designing a Verb P-controller

Chunk (Rule)	Statements of Verb Rules
$C1$	IF $e(k)$ stays zero THEN $\tilde{K}_p$ increases to big
$C2$	IF $e(k)$ decreases from big to medium THEN $\tilde{K}_p$ decreases fast from big
$C3$	IF $e(k)$ decreases from medium to small THEN $\tilde{K}_p$ decreases from big to small
$C4$	IF $e(k)$ decreases from small to zero THEN $\tilde{K}_p$ decreases slowly to small
$C5$	IF $e(k)$ increases from zero to small THEN $\tilde{K}_p$ increases slowly from small
$C6$	IF $e(k)$ increases from small to medium THEN $\tilde{K}_p$ increases from small to big
$C7$	IF $e(k)$ increases from medium THEN $\tilde{K}_p$ increases fast to big

Table 7.4. Canonical Forms in *become* of all Rules in Table 7.3

Chunk (Rule)	Statements of verb rules in canonical forms in *become*
$C1$	IF $|e(k)|$ *become*(zero, zero) THEN $\tilde{K}_p$ *become*(small, big)
$C2$	IF $|e(k)|$ *become*(big, medium) THEN $\tilde{K}_p$ *fast* ∘ *become*(big, small)
$C3$	IF $|e(k)|$ *become*(medium, small) THEN $\tilde{K}_p$ *become*(big, small)

C4	IF $\|e(k)\|$ *become*(small, zero) THEN $\tilde{K}_p$ slowly ∘ *become*(big, small)
C5	IF $\|e(k)\|$ *become*(zero, small) THEN $\tilde{K}_p$ slowly ∘ *become*(small, big)
C6	IF $\|e(k)\|$ *become*(small, medium) THEN $\tilde{K}_p$ *become*(small, big)
C7	IF $\|e(k)\|$ *become*(medium, big) THEN $\tilde{K}_p$ fast ∘ *become*(small, big)

D. Constructing Verb Rules for Tuning the Gain $\tilde{K}_d$ from Phase Plots

Observe that one can get the configuration shown in Figure 7.10(b) via rotating that in Figure 7.10(a) by 90-degree clockwise. Therefore, it is not necessary to construct verb rules from scratching. Instead, one can take advantage of the relation between both configurations shown in Figures 7.10(a) and 7.10(b) for the purpose of constructing a verb rule base for $\tilde{K}_d$. The verb rule base for tuning $\tilde{K}_d$ is simply a revision of the rule base shown in Tables 7.3 and 7.4. The resulting verb rule base for the D-controller is shown in Table 7.5.

Comparing the results shown in Tables 7.4 and 7.5, one can observe that for chunks C2 to C7 the corresponding verb rules can be obtained from their counterparts by exchanging the order of state 1 and state 2 in their conclusion parts. In doing so, the verb rule corresponding to chunk C1 remains unchanged.

Table 7.5. Canonical Forms in *become* of Verb Rules for Chunks $C1$ to $C7$ in
Figure 7.11 for Designing the Verb D-controller

Chunk (Rule)	Statements of verb rules in canonical forms in *become*
$C1$	IF $|e(k)|$ *become*(zero, zero) THEN $\tilde{K}_d$ *become*(small, big)
$C2$	IF $|e(k)|$ *become*(big, medium) THEN $\tilde{K}_d$ fast ∘ *become*(small, big)
$C3$	IF $|e(k)|$ *become*(medium, small) THEN $\tilde{K}_d$ *become*(small, big)
$C4$	IF $|e(k)|$ *become*(small, zero) THEN $\tilde{K}_d$ slowly ∘ *become*(small, big)
$C5$	IF $|e(k)|$ *become*(zero, small) THEN $\tilde{K}_d$ slowly ∘ *become*(big, small)
$C6$	IF $|e(k)|$ *become*(small, medium) THEN $\tilde{K}_d$ *become*(big, small)
$C7$	IF $|e(k)|$ *become*(medium, big) THEN $\tilde{K}_d$ fast ∘ *become*(big, small)

Table 7.6. Canonical Forms in *become* of Verb Rules for Chunks $C1$ to $C7$ in
Figure 7.11 for Designing the Verb I-controller

Chunk (Rule)	Statements of verb rules in canonical forms in *become*
$C1$	IF $|e(k)|$ *become*(zero, zero) THEN $\tilde{\gamma}$ *become*(small, big)
$C2$	IF $|e(k)|$ *become*(big, medium) THEN $\tilde{\gamma}$ fast ∘ *become*(small, big)
$C3$	IF $|e(k)|$ *become*(medium, small) THEN $\tilde{\gamma}$ *become*(small, big)
$C4$	IF $|e(k)|$ *become*(small, zero) THEN $\tilde{\gamma}$ slowly ∘ *become*(small, big)
$C5$	IF $|e(k)|$ *become*(zero, small) THEN $\tilde{\gamma}$ slowly ∘ *become*(big, small)
$C6$	IF $|e(k)|$ *become*(small, medium) THEN $\tilde{\gamma}$ *become*(big, small)
$C7$	IF $|e(k)|$ *become*(medium, big) THEN $\tilde{\gamma}$ fast ∘ *become*(big, small)

E. Constructing Verb Rules for Tuning $\tilde{\gamma}$ from Phase Plots

Comparing the configurations in Figures 6.10(b) and 6.10(c), one can observe that both configurations are similar if one treats the values that are bigger than 5 as PB and that less than 2 as PS. Here, the same membership functions are chosen for both $\tilde{K}_p$ and $\tilde{K}_d$ shown in Figure 7, which are used to fuzzify $\tilde{\gamma}$ as

$$\tilde{\gamma} = \frac{\gamma - \gamma_{\min}}{\gamma_{\max} - \gamma_{\min}}, \qquad\qquad (7.26)$$

where $\gamma_{\min} = 2$ and $\gamma_{\max} = 5$ in the present case. By doing so, the rule bases for $\tilde{K}_d$ and $\tilde{\gamma}$ are both in the same form. Therefore, one can simply modify the rule base for $\tilde{K}_d$ into that for $\tilde{\gamma}$, as listed in Table 7.6.

Since the design procedures for the P-, I- and D-controllers are the same, only the design process for the P-controller is presented and discussed below for brevity.

F.   Implementing the Verb-based P-controller

First, one needs to implement the seven rules listed in Table 7.4, namely, to implement a verb-based P-controller.

The reason for implementing the rules in Table 7.4 instead of those in Table 7.3 is that the rules in Table 7.4 are in more generic forms for which many design methods can be used. However, it should be remarked that the rules in Table 7.3 are only valid for designing the P-controller; therefore, the design methods used in designing Table 7.3 may not be easy to be reused in other cases such as in the design of the I- and D-controllers.

Choosing the Verb Sets

To implement the rule base shown in Table 7.4, one first needs to select two verb sets, V_x and V_y, or two adverb sets, A_x and A_y, so as to cover the entire *universe of dynamics* in interest.

Note that the universe of dynamics of a control system is roughly equivalent to all perceptible dynamics for the control system. Therefore, from the qualitative theory of dynamical systems, the universe of dynamics is a continuum of continuous-time dynamical systems. Recall that in general fuzzy controllers design, with a finite number of fuzzy sets, it is able to cover the entire universe of discourse. Similarly, in verb-based controllers design it is able to cover the entire universe of dynamics by using a finite number of verbs and adverbs.

More precisely, in the case of the verb-based P-controller, the universe of dynamics consists of the dynamics of the control error $e(t)$ and the normalized time-varying proportional gain $\tilde{K}_p(t)$.

Since the dynamics of $|e(t)|$ is chunked into four fuzzy values, *big*, *medium*, *small* and *zero*, it follows from the spiral trajectories shown in Figure 7.11 that the dynamics of $|e(t)|$ can be chunked into the following computational verbs:

Label	Computational Verbs for Dynamics of $|e(t)|$
V_{e1}	*become*(big, medium)
V_{e2}	*become*(medium, small)
V_{e3}	*become*(small, zero)
V_{e4}	*become*(zero, small)
V_{e5}	*become*(small, medium)
V_{e6}	*become*(medium, big)
V_{e7}	*become*(zero, zero)

On the other hand, since the dynamics of $\tilde{K}_p(t)$ can be chunked into two values, *small* and *big*, the dynamics of $\tilde{K}_p(t)$ can be chunked into the following computational verbs:

Label	Computational Verbs for Dynamics of $|e(t)|$
V_{p1}	*become*(big, small)
V_{p2}	*become*(small, big)

In summary, the most important issue in implementing computational verbs used in these rules is the implementation of the computational verb *become*.

Implementing the *become*(state 1, state 2) Verb Rule

As the generic canonical form of all verbs used in the verb rules, *become*(state 1, state 2) should be constructed based on the features of the dynamics of the underlying control system.

In the design of a verb-based PID controller, "state 1" and "state 2" are two implicit verbs of "being," namely, the initial and the final states of the controller. Assume that the dynamics of *become*(state 1, state 2) has an outer representation (i.e., measurement) $x(t)$, which can be either $|e(t)|$ or $\tilde{K}_p(t)$. The evolving function of *become*(state 1, state 2) can be defined based on the two associated membership functions $\mu_{\text{state 1}}(x(t))$ and $\mu_{\text{state 2}}(x(t))$. In doing so, one can restrict the range of $\varepsilon_{\text{become}}$ to be within $[0,1]$ and, therefore, avoid the trouble of coping with the amplitudes of individual membership functions on-line. Assume also that the life span of the computational verb is T_w. Then, the evolving function of *become*(state 1, state 2) can be defined as

$$\varepsilon_{\text{become(state 1, state 2)}}(t) = \begin{cases} \mu_{\text{state 1}}(x(t)), & t\in[0,T_w/2], \\ \mu_{\text{state 2}}(x(t)), & t\in[T_w/2,T_w]. \end{cases} \qquad (7.27)$$

This evolving function must satisfy the following criteria in order to be perceptually correct:

1. $\mu_{\text{state 1}}(x(0)) \geq \mu_{\text{state 1}}(x(T_w/2));$

2   $\mu_{\text{state 2}}(x(T_w/2)) \leq \mu_{\text{state 2}}(x(T_w));$

3. $\mu_{\text{state 1}}(x(0)) \approx 1$ and $\mu_{\text{state 2}}(x(T_w)) \approx 1.$

Observe that the values of $\mu_{\text{state 1}}(x(T_w/2))$ and $\mu_{\text{state 2}}(x(T_w/2))$ are not necessarily identical although they are usually chosen to be equal in order to reflect some "continuity" of the transient from state 1 to state 2. If one knows the parameters Tw, $\mu_{\text{state 1}}(x(T_w/2)) = a$, and $\mu_{\text{state 2}}(x(T_w/2)) = b$, then the simplest evolving function is given by

$$\varepsilon_{\text{become(state 1, state 2)}}(t) = \begin{cases} \dfrac{1}{T_w}(T_w - 2t + 2at), & t\in[0,T_w/2], \\ \dfrac{1}{T_w}(2t - T_w + 2b(T_w - t)), & t\in[T_w/2,T_w]. \end{cases} \qquad (7.28)$$

Calculating the Verb Similarities

Let the life span of *become*(state 1, state 2) be T_w, and an observed $x(t)$ be recorded. Then, one can construct the evolving function of a default observing verb, *observed*, based on the available $x(t)$, by

$$\varepsilon_{observed}(t) = \begin{cases} \mu_{\text{state 1}}(x(\tau)), & \tau \in [t - T_w, t - T_w / 2], \\ \mu_{\text{state 2}}(x(\tau)), & \tau \in [t - T_w / 2, t], \\ \text{undefined}, & \text{otherwise}, \end{cases} \tag{7.29}$$

where t denotes the current time instant.

Note that the life time for the observing verb *observed* is already in the past. Note also that the construction of the evolving function of the computational verb *observed* needs to take advantage of both the noun center and the verb center of their physical linguistics. More precisely, if one wants to know the verb similarity between *observed* and *become*(state 1, state 2), it requires to set the initial time of *become*(state 1, state 2) at the time instant $(t - T_w)$. In continuous time, the most commonly used verb similarities can be calculated by the following steps.

1.   The first half window:

$$a_1 = \int_0^{T_w/2} \varepsilon_{become}(\tau) \wedge \varepsilon_{observed}(t - T_w + \tau) d\tau, \tag{7.30a}$$

$$b_1 = \int_0^{T_w/2} \varepsilon_{become}(\tau) \vee \varepsilon_{observed}(t - T_w + \tau) d\tau. \tag{7.30b}$$

2.   The second half window:

$$a_2 = \int_{T_w/2}^{T_w} \varepsilon_{become}(\tau) \wedge \varepsilon_{observed}(t - T_w + \tau) d\tau, \tag{7.31a}$$

$$b_2 = \int_{T_w/2}^{T_w} \varepsilon_{become}(\tau) \vee \varepsilon_{observed}(t - T_w + \tau) d\tau. \tag{7.31b}$$

3. The balance factor ϖ:

$$\varpi = 2 \min\left(\frac{a_1}{b_1 + b_2}, \frac{a_2}{b_1 + b_2}\right). \tag{7.32}$$

4. The entire window:

$$S(become, observed) = \left(\frac{a_1 + a_2}{b_1 + b_2}\right)\varpi. \tag{7.33}$$

For discrete time, the verb similarity can be calculated by using samples of the evolving functions. Note that one can calculate the verb similarities at every point of a measured time series $x(t)$, $t = 0, 1, 2, \ldots$, by using the evolving function *become* as a moving template over a window of fixed length, T_w. This is the basic technique that will be used to control a system along time, as further discussed below.

G. Implementing the Verb-based P-controller for a Second-Order Plant

Consider the following plant as an example for illustrating the design process of a verb-based P-controller:

$$\dot{x}_1(t) = x_2(t), \tag{7.34a}$$

$$\dot{x}_2(t) = -2x_2(t) - x_1(t) + u(t-T_s), \tag{7.34b}$$

$$y(t) = x_1(t), \tag{7.34c}$$

$$e(t) = r - y(t), \tag{7.34d}$$

where $x_i(t)$, $i = 1, 2$, are two state variables of the plant, $y(t)$ is the output, $u(t)$ is the input control signal, r is the reference, and T_s is the sampling period.

Observe that in model (7.34) the control is delayed, caused by the time period that is needed to calculate the control signal.

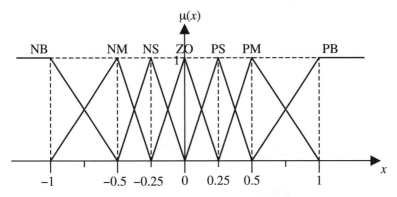

Figure 6.12. Membership functions of fuzzy numbers splitting the normalized range of control error $e(k)$

Fuzzy Sets

For practical considerations, the controlled plant must be asymptotically stable, or at least bounded-input bounded-output stable. It is safe to consider errors in the range of $[-|e(0)|, |e(0)|]$ with a large value of $|e(0)|$.

Traditionally, one uses membership functions as those shown in Figure 7.12 to equally divide the error value range. However, since $e(t)$ will exponentially approach zero in an asymptotically stable controlled system, one has to pay more attention to the region closed to zero because trajectories stay there longer than in other regions far away from zero. Based on this consideration, the supports of the membership functions can be chosen as that shown in Figure 7.12, where the range for the tracking error has been normalized. The boundaries of regions shown in Figure 7.10 use the supports of the fuzzy sets shown in Figure 7.12. These membership functions are given as follows:

$$\mu_{PB}(x) = \max(\,0,\, 1 + (x-1) - |x-1|\,), \tag{7.35a}$$

$$\mu_{PM}(x) = \max(\,0,\, 1 + (x-0.5) - 3|x-0.5|\,), \tag{7.35b}$$

$$\mu_{PS}(x) = \max(\,0,\, 1 - 4|x-0.25|\,), \tag{7.35c}$$

$$\mu_{ZO}(x) = \max(\,0,\, 1 - 4|x|\,), \tag{7.35d}$$

$$\mu_{NS}(x) = \max(\,0,\, 1 - 4|x+0.25|\,), \tag{7.35e}$$

$$\mu_{NM}(x) = \max(\,0,\, 1 - (x+0.5) - 3|x+0.5|\,), \tag{7.35f}$$

$$\mu_{NB}(x) = \max(\ 0,\ 1 - (x+1) - |x+1|\ ).\qquad\qquad(7.35g)$$

Although the membership functions for $\tilde{K}_p$ and $\tilde{K}_d$ shown in Figure 7.7 may not be optimal, for the purpose of comparison, they are used here without modifications.

Become Functions

Analytic representations of all *become* functions used in computational verbs V_{e1} to V_{e7}, V_{p1} and V_{p2} are first introduced.

For the purpose of generic implementation, all *become* functions are chosen to be independent of any specific dynamics such that one can program a "universal controller" with as fewer parameters left for the users to adjust as possible. To do so, the evolving function of every *become* is chosen as the piecewise linear function (7.28) with $a = b = 0.5$ therein; namely,

$$\varepsilon_{become(state\ 1,\ state\ 2)} = \begin{cases} \dfrac{T_w - t}{T_w}, & t \in [0, T_w / 2], \\[3mm] \dfrac{t}{T_w}, & t \in [T_w / 2, T_w]. \end{cases}\qquad(7.36)$$

Its discrete-time version is given by

$$\varepsilon_{become(state\ 1,\ state\ 2)} = \begin{cases} \dfrac{w - k}{w}, & k \in [0, w / 2],\ w \in Z, \\[3mm] \dfrac{k}{w}, & k \in (w / 2, w],\ k \in Z. \end{cases}\qquad(7.37)$$

Note that in the discrete case, the life span of *become* contains $(w+1)$ sampling points. If w is even then the first half of the window will be 1 sampling point longer than the second half of the window; if w is odd then both half-windows have the same length.

Verb Similarity Functions

Before one can perform any verb inference, it is needed to find verb similarity functions of all verbs in the following verb sets:

$$V_e = \{\ V_{e1},\ V_{e2},\ V_{e3},\ V_{e4},\ V_{e5},\ V_{e6},\ V_{e7}\ \},$$

$$V_p = \{\ V_{p1},\ V_{p2}\ \}.$$

Since one needs to know the evolving functions of computational verbs before one can calculate a verb similarity function, it is convenient to first find the evolving functions and the evolving processes of $e(t)$ and $\tilde{K}_p(t)$ for all verbs in V_e and V_p. This can be done in the following steps.

1. $V_{e1} = become$(big, medium). The evolving function is given by

$$\varepsilon_{become} = \begin{cases} \mu_{PB}(|e(t)|) = \dfrac{T_w - t}{T_w}, & t \in [0, T_w/2], \\[4mm] \mu_{PM}(|e(t)|) = \dfrac{t}{T_w}, & t \in [T_w/2, T_w]. \end{cases} \tag{7.38}$$

Note that the corresponding time series of $|e(t)|$, $e_1(t)$, passes points $(0,1)$, $(T_w/2, 0.75)$ and $(T_w, 0.5)$. Therefore, there is linear interpretation of V_{e1} as

$$e_1(t) = 1 - \frac{1}{2T_w}, \qquad t \in [0, T_w]. \tag{7.39}$$

2. $V_{e2} = become$(medium, small). The evolving function is given by

$$\varepsilon_{become} = \begin{cases} \mu_{PM}(|e(t)|) = \dfrac{T_w - t}{T_w}, & t \in [0, T_w/2], \\[4mm] \mu_{PS}(|e(t)|) = \dfrac{t}{T_w}, & t \in [T_w/2, T_w]. \end{cases} \tag{7.40}$$

3. $V_{e3} = become$(small, zero). The evolving function is given by

$$\varepsilon_{become} = \begin{cases} \mu_{PS}(|e(t)|) = \dfrac{T_w - t}{T_w}, & t \in [0, T_w/2], \\[4mm] \mu_{ZO}(|e(t)|) = \dfrac{t}{T_w}, & t \in [T_w/2, T_w]. \end{cases} \tag{7.41}$$

4. $V_{e4} = become$(zero, small). The evolving function is given by

$$\varepsilon_{become} = \begin{cases} \mu_{ZO}(|e(t)|) = \dfrac{T_w - t}{T_w}, & t \in [0, T_w/2], \\[4mm] \mu_{PS}(|e(t)|) = \dfrac{t}{T_w}, & t \in [T_w/2, T_w]. \end{cases} \tag{7.42}$$

5. $V_{e5} = become$(small, medium). The evolving function is given by

$$
\varepsilon_{become} = \begin{cases} \mu_{PS}(|e(t)|) = \dfrac{T_w - t}{T_w}, & t \in [0, T_w/2], \\[2mm] \mu_{PM}(|e(t)|) = \dfrac{t}{T_w}, & t \in [T_w/2, T_w]. \end{cases} \tag{7.43}
$$

6. $V_{e6} = become$(medium, big). The evolving function is given by

$$
\varepsilon_{become} = \begin{cases} \mu_{PM}(|e(t)|) = \dfrac{T_w - t}{T_w}, & t \in [0, T_w/2], \\[2mm] \mu_{PB}(|e(t)|) = \dfrac{t}{T_w}, & t \in [T_w/2, T_w]. \end{cases} \tag{7.44}
$$

7. $V_{e7} = become$(zero, zero). The evolving function is given by

$$
\varepsilon_{become} = \begin{cases} \mu_{ZO}(|e(t)|) = \dfrac{T_w - t}{T_w}, & t \in [0, T_w/2], \\[2mm] \mu_{ZO}(|e(t)|) = \dfrac{t}{T_w}, & t \in [T_w/2, T_w]. \end{cases} \tag{7.45}
$$

Next, the notation V_1 (V_2) is used is used to denote the *degree of similarity* of verb V_2 to the *reference verb* V_1, and the verb similarity function of verb V with respect to a verb set V is denoted by

$$
\xi_V^V = \sum_{V_x \in V} \frac{\xi_V(V_x)}{V_x}. \tag{7.46}
$$

Assume that the canonical forms in *become* of V_1 and V_2 are *become*(state 1a, state 2a) and *become*(state 1b, state 2b), respectively. Moreover, let $x_1(t)$ be an interpretation of V_1 in the form of

$$
x_1(t) = \begin{cases} f(t), & t \in [0, T_w/2], \\ g(t), & t \in [T_w/2, T_w]. \end{cases} \tag{7.47}
$$

The degree of similarity $\xi_{V_1}(V_2)$ can be calculated in the following steps.

1. Calculate $a_1, a_2, b_1,$ and b_2:

$$a_1 = \int_0^{T_w/2} \min(\mu_{\text{state 1a}}(f(t)), \mu_{\text{state 1b}}(f(t)))dt \,, \qquad (7.48a)$$

$$a_2 = \int_{T_w/2}^{T_w} \min(\mu_{\text{state 2a}}(g(t)), \mu_{\text{state 2b}}(g(t)))dt \,, \qquad (7.48b)$$

$$b_1 = \int_0^{T_w/2} \max(\mu_{\text{state 1a}}(f(t)), \mu_{\text{state 1b}}(f(t)))dt \,, \qquad (7.48c)$$

$$b_2 = \int_{T_w/2}^{T_w} \max(\mu_{\text{state 2a}}(g(t)), \mu_{\text{state 2b}}(g(t)))dt \,. \qquad (7.48d)$$

2. Calculate the balance factor ϖ:

$$\varpi = 2 \min\left(\frac{a_1}{b_1+b_2}, \frac{a_2}{b_1+b_2}\right). \qquad (7.49)$$

3. Calculate the degree of similarity:

$$\xi_{V_1}(V_2) = \frac{a_1+a_2}{b_1+b_2}\varpi. \qquad (7.50)$$

The verb similarity functions of all verbs in V_e can be easily calculated as shown by the following example, which shows how $\xi_{V_{e1}}^{V_e}$ is calculated.

Example 7.2 $\left(\text{Find } \xi_{V_{e1}}^{V_e}\right)$

As an illustrating example, only $\xi_{V_1}(V_{e2})$ is calculated here.

The linear interpretation shown in (7.39) is used for the verb V_{e1}. It follows from (7.35), (7.39), and (7.48) that

$$a_1 = \int_0^{T_w/2} \min(\mu_{PB}(e_1(t)), \mu_{PM}(e_1(t)))dt$$

$$= \int_0^{T_w/2} \mu_{PM}\left(1 - \frac{t}{2T_w}\right) dt$$

$$= \int_0^{T_w/2} \max\left(0, 1 + \left[\left(1 - \frac{t}{2T_w}\right) - 0.5\right] - 3\left|\left(1 - \frac{t}{2T_w}\right) - 0.5\right|\right) dt$$

$$= \int_0^{T_w/2} \frac{t}{T_w} dt$$

$$= \frac{T_w}{8}, \tag{7.51}$$

$$a_2 = \int_{T_w/2}^{T_w} \min(\mu_{PM}(e_1(t)), \mu_{PS}(e_1(t))) dt$$

$$= \int_{T_w/2}^{T_w} \mu_{PS}\left(1 - \frac{t}{2T_w}\right) dt$$

$$= \int_{T_w/2}^{T_w} \max\left(0, 1 - 4\left|\left(1 - \frac{t}{2T_w}\right) - 0.25\right|\right) dt$$

$$= 0, \tag{7.52}$$

$$b_1 = \int_0^{T_w/2} \max(\mu_{PB}(e_1(t)), \mu_{PS}(e_1(t))) dt$$

$$= \int_0^{T_w/2} \mu_{PB}\left(1 - \frac{t}{2T_w}\right) dt$$

$$= \int_0^{T_w/2} \max\left(0, 1 + \left[\left(1 - \frac{t}{2T_w}\right) - 1\right] - \left|\left(1 - \frac{t}{2T_w}\right) - 1\right|\right) dt$$

$$= \int_0^{T_w/2} \left(1 - \frac{t}{T_w}\right) dt$$

$$= \frac{3T_w}{8}, \tag{7.53}$$

$$b_2 = \int_{T_w/2}^{T_w} \max(\mu_{PM}(e_1(t)), \mu_{PS}(e_1(t))) dt$$

$$= \int_{T_w/2}^{T_w} \mu_{PM} \left(1 - \frac{t}{2T_w}\right) dt$$

$$= \int_{T_w/2}^{T_w} \max\left(0, 1 + \left[\left(1 - \frac{t}{2T_w}\right) - 0.5\right] - 3\left|\left(1 - \frac{t}{2T_w}\right) - 0.5\right|\right) dt$$

$$= \int_{T_w/2}^{T_w} \frac{t}{T_w} dt$$

$$= \frac{3T_w}{8}. \tag{7.54}$$

Therefore, one has $\varpi = 0$ and $\xi_{V_{e1}}^{V_e}(V_{e2}) = 0$. Using the same method, one can find other elements of $\xi_{V_{e1}}^{V_e}$ as

$$\xi_{V_{e1}}^{V_e} = \frac{1}{V_{e1}} + \frac{0}{V_{e2}} + \frac{0}{V_{e3}} + \frac{0}{V_{e4}} + \frac{0}{V_{e5}} + \frac{1/9}{V_{e6}} + \frac{0}{V_{e7}}.$$

Assume a linear interpretation of $\tilde{K}_p(t)$. Then, the verb similarity functions of both verbs in V_p can be found as follows.

1. Due to the symmetry of μ_{PB} and μ_{PS} for $\tilde{K}_p(t)$, the quantity $\xi_{V_{p1}}^{V_p}$ can be found by the following processes:

$$a_1 = \int_{0.5}^{1} -\frac{1}{4} \ln(x) dx$$

$$= \left. -\frac{x}{4}\ln(x) + \frac{x}{4} \right|_{0.5}^{1}$$

$$= 0.0384, \tag{7.55a}$$

$$b_1 = e^{-4} + \int_{e^{-4}}^{0.5} -\frac{1}{4}\ln(x)dx$$

$$= \left. e^{-4} - \frac{x}{4}\ln(x) + \frac{x}{4} \right|_{e^{-4}}^{0.5}$$

$$= 0.207, \tag{7.55b}$$

$$a_2 = 0.0384, \tag{7.55c}$$

$$b_2 = 0.207, \tag{7.55d}$$

$$\varpi = 2 \times \frac{0.0384}{2 \times 0.207}$$

$$= 0.1855, \tag{7.55e}$$

$$\xi_{V_{p1}}^{V_{p2}} = \varpi \frac{a_1 + a_2}{b_1 + b_2}$$

$$= 0.1855 \times \frac{0.0384}{0.207}$$

$$= 0.0344. \tag{7.55f}$$

Here,

$$\xi_{V_{p1}}^{V_p} = \frac{1}{V_{p1}} + \frac{0.0344}{V_{p2}}.$$

2. From the symmetry of μ_{PB} and μ_{PS}, one has

$$\xi_{V_{p2}}^{V_p} = \frac{0.0344}{V_{p1}} + \frac{1}{V_{p2}}.$$

Since 0.0344 is quite small as compared to 1, in practical applications, one may ignore it.

Verb Inferences

With all verb similarity functions defined and found, one can now proceed to implement the verb inferences based on the verb rules shown in Table 4.

For each rule, one first finds an adverb α, called *firing adverb*, by comparing the verb similarity between the verb phrase in its condition part and the computational verb observed, which is the verbifying result of $|e(k)|$. By applying the adverb to the verb phrase in the conclusion part of the verb rule, one can calculate the contribution of this rule to the change of $\tilde{K}_p$. The final decision on changing $\tilde{K}_p$ is found by applying a de-verbification to the contributions from all the rules. Here, the following de-verbification formula is used:

$$K_p(k+1) = K_p(k) + \delta \sum_{i=0}^{7} \alpha_i \xi_i \, , \tag{7.56}$$

where α_i and ξ_i are the firing adverb and full contribution from rule i, respectively, and δ is a weighting factor. The de-verbifying results of V_{p1} and V_{p2} are denoted by ξ_{p1} and ξ_{p2}, respectively, and are given by

$$\xi_{p1} = e^{\{10[S(observed, V_{p2}) - S(observed, V_{p1})]\}}, \tag{7.57a}$$

$$\xi_{p2} = e^{\{10[S(observed, V_{p1}) - S(observed, V_{p2})]\}}. \tag{7.57b}$$

Observe that ξ_{p1} and ξ_{p2} are two data for the de-verbification of V_{p1} and $V_{p2.}$ The intuition behind formula (7.57) is that the dynamics of $K_p(t)$ are spanned by two complementary verbs; namely, V_{p1} and V_{p2}.

Assume that one wants the dynamics of $K_p(t)$ to be the same as V_{p1}. Then, there are two cases to consider.

1. If, at a moment, one observes that $K_p(t)$ is more like V_{p1}. Then, one only needs to change it slowly because it behaves as expected. This is implemented by setting $\xi_{p1} \le 1$.

2. If, at a moment, one observes that $K_p(t)$ is more like V_{p2}. Then, one needs to change it into V_{p1} in a fast speed because one is expecting $V_{p1.}$ and this is implemented by setting $\xi_{p1} > 1$.

In the above rules, two adverbs (fast and slowly) are used to modify the dynamics of V_{p1} and V_{p2}. The relation between the de-verbification of V_{p1} and $fast \circ V_{p1}$ is given by

$$\xi_{fast \circ V_{p1}} = \beta_{fast} \times \xi_{p1}. \tag{7.58}$$

Also, the relation between the de-verbification of V_{p1} and $slowly \circ V_{p1}$ is given by

$$\xi_{slowly \circ V_{p1}} = \beta_{slowly} \times \xi_{p1}. \tag{7.59}$$

The same is applied to V_{p2} with two adverbs, *fast* and *slowly*.

Now, one can calculate the contributions from all rules as follows:

1. Implement the first rule:

    C1:   IF $|e(k)|$ *become*(zero, zero)
            THEN $\tilde{K}_p$ *become*(small, big).

    The contribution of this rule is

    $$\Lambda(\alpha_1 \circ [S(observed, become(small, big))]) = \alpha_1 \times \xi_{e2},$$

    where $\Lambda(\cdot)$ denotes the effect of the de-verbifying block.

2. Implement the second rule:

    C2:   IF $|e(k)|$ *become*(big, medium)
            THEN $\tilde{K}_p$ *fast* $\circ$ *become*(big, small).

    The contribution of this rule is

    $$\Lambda(\alpha_2 \circ [S(observed, fast \circ become(big, small))])$$

    $$= -\alpha_2 \times \xi_{fast \circ V_{p1}}$$

    $$= -\alpha_2 \times \beta_{fast} \times \xi_{p1},$$

    where the negative sign reflects the fact that $\tilde{K}_p$ decreases with a negative changing rate.

3. Implement the third rule:

$C3$: IF $|e(k)|$ *become*(medium, small)
THEN $\tilde{K}_p$ *become*(big, small).

The contribution of this rule is

$$\Lambda(\alpha_3 \circ [S(observed, become(\text{big, small}))]) = -\alpha_3 \times \xi_{p1}.$$

4. Implement the fourth rule:

$C4$: IF $|e(k)|$ *become*(small, zero)
THEN $\tilde{K}_p$ *slowly* $\circ$ *become*(big, small).

The contribution of this rule is

$$\Lambda(\alpha_4 \circ [S(observed, slowly \circ become(\text{big, small}))])$$

$$= -\alpha_4 \times \xi_{slowly \circ v_{p1}}$$

$$= -\alpha_4 \times \beta_{slowly} \times \xi_{p1}.$$

5. Implement the fifth rule:

$C5$: IF $|e(k)|$ *become*(zero, small)
THEN $\tilde{K}_p$ *slowly* $\circ$ *become*(small, big).

The contribution of this rule is

$$\Lambda(\alpha_5 \circ [S(observed, slowly \circ become(\text{small, big}))])$$

$$= \alpha_5 \times \xi_{slowly \circ v_{p2}}$$

$$= \alpha_5 \times \beta_{slowly} \times \xi_{p2}.$$

6. Implement the sixth rule:

$C6$: IF $|e(k)|$ *become*(small, medium)
THEN $\tilde{K}_p$ *become*(small, big).

The contribution of this rule is

$$\Lambda(\alpha_6 \circ [S(observed, become(\text{small, big}))]) = \alpha_6 \times \xi_{p2}.$$

7. Implement the seventh rule:

$C7$: IF $|e(k)|$ *become*(medium, big),

$$\text{THEN } \tilde{K}_p \, fast \circ become(\text{small, big}).$$

The contribution of this rule is

$$\Lambda(\alpha_7 \circ [S(observed, fast \circ become(\text{small, big}))])$$

$$= \alpha_7 \times \tilde{\xi}_{fast \circ v_{p2}}$$

$$= \alpha_7 \times \beta_{fast} \times \tilde{\xi}_{p2}.$$

Simulation Results

The simulation results are shown in Figure 7.13.

In Figure 7.13(a), the first row shows the curve of $e(t)$. In the second row, $\mu_B(e(t))$, $\mu_M(e(t))$, $\mu_S(e(t))$, and $\mu_Z(e(t))$ are shown by solid, dashed, dash-dotted, and dotted curves, respectively. In the third row, the verb similarities between $observed \, |e(k)|$ and $|e(k)| \, become(\text{zero, zero})$, $become(\text{big, medium})$, $become(\text{medium, small})$ are shown, respectively, by solid, dashed and dash-dotted curves. In the fourth row, the verb similarities between $observed \, |e(k)|$ and $|e(k)| \, become(\text{small, zero})$, $become(\text{zero, small})$, $become(\text{small, medium})$ are shown, respectively, in solid, dashed and dash-dotted curves. Since the verb similarity between $observed \, |e(k)|$ and $|e(k)| \, become(\text{medium, big})$ is zero everywhere, it is not shown in the figure. In the fifth row, $K_p(k)$ is shown by a solid curve, while $\mu_{big}(\tilde{K}_p(k))$ and $\mu_{small}(\tilde{K}_p(k))$ are shown by dashed-dotted and dash curves, respectively. The amplitudes of the curves of $\mu_{big}(\tilde{K}_p(k))$ and $\mu_{small}(\tilde{K}_p(k))$ are enlarged 15 times for the sake of visibility. The verb similarities between $observed \, K_p(k)$ and $\tilde{K}_p(k) \, become(\text{big, small})$, $become(\text{small, big})$ are shown, respectively, by dash-dotted and dashed curves. For the purpose of visibility, the amplitudes of the dash-dotted curves have been enlarged 200 times.

The parameters for the simulation are $\delta = 0.4$ (see (7.56)), $\beta_{fast} = 5$, and $\beta_{slowly} = 0.125$.

Figure 7.13(b) shows the firing adverbs $\alpha_1(t)$, ..., $\alpha_7(t)$ for all verb rules in the verb inference process. For comparing purpose, the curve of $K_p(t)$ is plotted by dotted curve in each row of this figure.

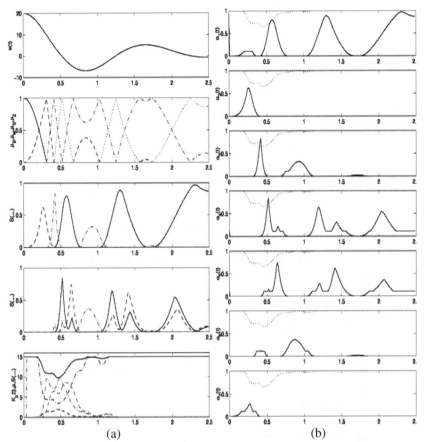

(a)   (b)

Figure 7.13. Control the 2nd-order system (7.34) using the verb-based
P-controller: (a) Control signal. (b) The evolutions of
firing adverbs, $\alpha_1(t)$, ..., $\alpha_7(t)$, of the rules in the verb
inference for the verb-based P-controller

Problems

P7.1   Find the evolving function of the following computational verb:

An apple *falls* down from the tree.

P7.2   Given the evolving function of the computational verb *decrease* as

$$\varepsilon_{decrease}(t) = h_0 + ke^{-t}, t \in [0, T_w],$$

where $h_0 \in R$ and $k \geq 0$ are two parameters to be determined, find the evolving function of the following computational verbs:

(i)   The temperature *decreases* from 10 degree to 0.

(ii)   The temperature *decreases* from the value around 10 degree to the value around 0.

P7.3   Given the evolving function of *increase* be

$$\varepsilon_{decrease}(t) = e^t,$$

find the evolving function of the following computational verbs:

(i)   *fast increase*

(ii)   *slowly increase*

(iii)   *decrease*

P7.4   Find the verb similarity between the following computational verbs:

(i) $\varepsilon_1 = e^{-t}$ and $\varepsilon_2 = e^{-2t}, t \in [0,2]$

(ii) $\varepsilon_1 = e^{-t}$ and $\varepsilon_2 = e^t, t \in [0,2]$

P7.5   Find the canonical computational verbs *become* of the following verbs:

(i)   *go* to the museum

(ii)   *speed* up

P7.6   Find the linear interpretations of Ve2 to Ve7, of which the evolving functions are given in the following equations:

(i) $V_{e2} = become$(medium, small), where the evolving function is given by

$$
\varepsilon_{become}(t) = \begin{cases} \mu_{PM}(|e(t)|) = \dfrac{T_w - t}{T_w}, & t \in [0, T_w / 2], \\[3mm] \mu_{PS}(|e(t)|) = \dfrac{t}{T_w}, & t \in [T_w / 2, T_w]. \end{cases}
$$

(ii) $V_{e3} = become$(small, zero), where the evolving function is given by

$$
\varepsilon_{become}(t) = \begin{cases} \mu_{PS}(|e(t)|) = \dfrac{T_w - t}{T_w}, & t \in [0, T_w / 2], \\[3mm] \mu_{ZO}(|e(t)|) = \dfrac{t}{T_w}, & t \in [T_w / 2, T_w]. \end{cases}
$$

(iii) $V_{e4} = become$(zero, small), where the evolve function is given by

$$
\varepsilon_{become}(t) = \begin{cases} \mu_{ZO}(|e(t)|) = \dfrac{T_w - t}{T_w}, & t \in [0, T_w / 2], \\[3mm] \mu_{PS}(|e(t)|) = \dfrac{t}{T_w}, & t \in [T_w / 2, T_w]. \end{cases}
$$

(iv) $V_{e5} = become$(small, medium), where the evolving function is given by

$$
\varepsilon_{become}(t) = \begin{cases} \mu_{PS}(|e(t)|) = \dfrac{T_w - t}{T_w}, & t \in [0, T_w / 2], \\[3mm] \mu_{PM}(|e(t)|) = \dfrac{t}{T_w}, & t \in [T_w / 2, T_w]. \end{cases}
$$

(v) $V_{e6} = become$(medium, big), where the evolving function is given by

$$\varepsilon_{become}(t) = \begin{cases} \mu_{PM}(|e(t)|) = \dfrac{T_w - t}{T_w}, & t \in [0, T_w/2], \\ \\ \mu_{PB}(|e(t)|) = \dfrac{t}{T_w}, & t \in [T_w/2, T_w]. \end{cases}$$

(vi) $V_{e7} = become(\text{zero, zero})$, where the evolving function is given by

$$\varepsilon_{become}(t) = \begin{cases} \mu_{ZO}(|e(t)|) = \dfrac{T_w - t}{T_w}, & t \in [0, T_w/2], \\ \\ \mu_{ZO}(|e(t)|) = \dfrac{t}{T_w}, & t \in [T_w/2, T_w]. \end{cases}$$

P7.7   Find the verb similarities $\xi^{V_e}_{V_{e2}}, \ldots, \xi^{V_e}_{V_{e7}}$.

[*Hint*] Use the linear interpretations of computational verbs V_{e2} to V_{e7} found in *P7.6* above.

REFERENCES

[1] G. Alefeld, and J. Herzberger. *Introduction to Interval Computations.* New York, NY: Academic Press (1983).

[2] C. R. Asfahl. *Robots and Manufacturing Automation.* New York, NY: Wiley (1992).

[3] O. Calvo, and J. H. E. Cartwright. "Fuzzy Control of Chaos," *Int. J. of Bifurcation and Chaos*, Vol. 8, No. 8, pp. 1743-1747 (1998).

[4] S. G. Cao, N. W. Rees, and G. Feng. "Stability Analysis of Fuzzy Control Systems." *IEEE Transactions on Systems, Man, and Cybernetics* (B), Vol. 26, No. 1, pp 201-204 (1996).

[5] G. Chen, and T. T. Pham. *Introduction to Fuzzy Sets, Fuzzy Logic, and Fuzzy Systems.* Boca Raton, FL: CRC Press (2001).

[6] G. Chen, T. T. Pham, and J. J. Weiss. "Fuzzy Modeling of Control System." *IEEE Transactions on Aerospace and Electronics Systems*, Vol. 31, No. 1, pp 414-429 (1995).

[7] C. K. Chui, and G. Chen. *Linear Systems and Optimal Control.* New York, NY: Springer-Verlag (1989).

[8] A. Deif. *Sensitivity Analysis in Linear Systems.* New York, NY: Springer-Verlag (1992).

[9] D. Drianker, H. Hellendoorn, and M. Reinfrank. *An Introduction to Fuzzy Control.* Berlin, Germany: Springer-Verlag (1993).

[10] E. Hansen. "Interval Arithmetic in Matrix Computations, Part I." *SIAM Journal of Numerical Analysis*, Vol. 2, pp. 308-320 (1965).

[11] E. Hansen. "On Linear Algebraic Equations with Interval Coefficients." *Topics in Interval Analysis* (E. Hansen, ed.). Oxford, England: Claredon Press (1969).

[12] E. Hansen and R. Smith. "Interval Arithmetic in Matrix Computations, Part II." *SIAM Journal of Numerical Analysis*, Vol. 4, pp. 1-9 (1967).

[13] A. Kaufmann and M. M. Gupta. *Introduction to Fuzzy Arithmetic: Theory and Applications.* New York, NY: Van Nostrand Reinhold (1991).

[14] R. B. Kearfott. *Rigorous Global Search: Continuous Problems.* Boston, MA: Kluwer Academic Publishers (1996).

[15] G. J. Klir, U. H. St. Clair, and B. Yuan. *Fuzzy Set Theory: Foundations and Applications.* Upper Saddle River, NJ: Prentice-Hall PTR (1997).

[16] G. J. Klir, and T. A. Folger. *Fuzzy Sets, Uncertainty, and Information.* Englewood Cliffs, NJ: Prentice-Hall (1988).

[17] B. Kosko. *Fuzzy Engineering.* Upper Saddle River, NJ: Prentice-Hall (1997).

[18] R. E. Moore. *Methods and Applications of Interval Analysis.* Philadelphia, PA: SIAM Press (1979).

[19]   K. M. Passino, and S. Yurkovich, *Fuzzy Control.* Menlo Park, CA: Addison-Wesley (1998).

[20]   K. Tanaka. "Design of model-based fuzzy control using Lyapunov's stability approach and its applications to trajectory stabilization of a model car." *Theoretical Aspects of Fuzzy Control* (ed. H. T. Nguyen, M. Sugeno, R. Tong, and R. R. Yager), pp 31-63, New York, NY: Wiley (1995).

[21]   K. Tanaka. *An Introduction to Fuzzy Logic for Practical Applications.* New York, NY: Springer (1996).

[22]   T. Takagi, and M. Sugeno. "Fuzzy Identification of Systems and Its Applications to Modeling and Control." *IEEE Transactions on Systems, Man, and Cybernetics*, Vol. 15, No.1, pp 116-132 (1985).

[23]   K. Tanaka, and M. Sugeno. "Stability Analysis and Design of Fuzzy Control Systems." *Fuzzy Sets and Systems*, Vol. 45, pp 135-156 (1992).

[24]   K. Tanaka, and H. O. Wang. *Fuzzy Control Systems Design and Analysis: A Linear Matrix Inequality Approach.* New York, NY: IEEE Press (1999).

[25]   L. X. Wang. *Adaptive Fuzzy Systems and Control: Design and Stability Analysis.* Englewood Cliffs, NJ: Prentice-Hall (1994).

[26]   J. W. Webb and R. A. Reis. *Programmable Logic Controllers: Principles and Applications.* Englewood Cliffs, NJ: Prentice-Hall (1995).

[27]   T. Yang. *Computational Verb Theory.* Tucson, AZ: Yang's Scientific Research Institute, LLC (2002).

[28]   H. Ying. *Fuzzy Control and Modeling: Analytical Foundations and Applications.* New York, NY: IEEE Press (2000).

[29]   L. A. Zadeh. "The role of fuzzy logic in modeling, identification, and control." *Modeling, Identification, and Control*, Vol. 15, No. 3, pp 191-203 (1994).

[30]   H. J. Zimmermann. *Fuzzy Set Theory and Its Applications (2nd Edition).* Boston, MA: Kluwer Academic Publishers (1991).

Solutions to Problems

CHAPTER 1

P1.1   Solution.

Consider the following four sets:

$$C = \{c \mid c \in A \cap B\}$$
$$D = \{d \mid d \in A \cap \bar{B}\}$$
$$E = \{e \mid e \in \bar{A} \cap B\}$$
$$F = \{f \mid f \in \bar{A} \cap \bar{B}\}$$

	$X_A(x)$	$X_B(x)$	$\max\{X_A(x), X_B(x)\}$	$\min\{X_A(x), X_B(x)\}$	$1 - X_A(x)$
$x = c$	1	1	1	1	0
$x = d$	1	0	1	0	0
$x = e$	0	1	1	0	1
$x = f$	0	0	0	0	1

It follows that

$$\max\{X_A(x), X_B(x)\} = \begin{cases} 1 & if \quad x \in (A \cap B) \cup (A \cap \bar{B}) \cup (\bar{A} \cap B) \\ 0 & if \quad x \in \bar{A} \cap \bar{B} \end{cases}$$

$$\begin{aligned} \because (A \cap B) \cup (A \cap \bar{B}) \cup (\bar{A} \cap B) &= (A \cap (B \cup \bar{B})) \cup (\bar{A} \cap B) \\ &= A \cup (\bar{A} \cap B) \\ &= (A \cup \bar{A}) \cap (A \cup B) \\ &= A \cup B \end{aligned}$$

$$\bar{A} \cap \bar{B} = \overline{A \cup B}$$

$$\therefore \max\{X_A(x), X_B(x)\} = \begin{cases} 1 & if \quad x \in A \cup B \\ 0 & if \quad x \in \overline{A \cup B} \end{cases}$$

$$= \begin{cases} 1 & if \quad x \in A \cup B \\ 0 & if \quad x \notin A \cup B \end{cases}$$

$$= X_{A \cup B}(x)$$

$$\min\{X_A(x), X_B(x)\} = \begin{cases} 1 & if \quad x \in A \cap B \\ 0 & if \quad x \in (A \cap \overline{B}) \cup (\overline{A} \cap B) \cup (\overline{A} \cap \overline{B}) \end{cases}$$

$$\because (A \cap \overline{B}) \cup (\overline{A} \cap B) \cup (\overline{A} \cap \overline{B}) = (A \cap \overline{B}) \cup (\overline{A} \cap (B \cup \overline{B}))$$

$$= (A \cap \overline{B}) \cup \overline{A}$$

$$= (A \cup \overline{A}) \cap (\overline{B} \cup \overline{A})$$

$$= \overline{B} \cup \overline{A}$$

$$= \overline{A} \cup \overline{B}$$

$$= \overline{A \cap B}$$

$$\therefore \min\{X_A(x), X_B(x)\} = \begin{cases} 1 & if \quad x \in A \cap B \\ 0 & if \quad x \in \overline{A \cap B} \end{cases}$$

$$= \begin{cases} 1 & if \quad x \in A \cap B \\ 0 & if \quad x \notin A \cap B \end{cases}$$

$$= X_{A \cap B}(x)$$

$$1 - X_A(x) = \begin{cases} 1 & if \quad x \in (\overline{A} \cap B) \cup (\overline{A} \cap \overline{B}) \\ 0 & if \quad x \in (A \cap B) \cup (A \cap \overline{B}) \end{cases}$$

$$\because (\overline{A} \cap B) \cup (\overline{A} \cap B) = \overline{A} \cap (B \cup \overline{B})$$

$$= \overline{A}$$

$$(A \cap B) \cup (A \cap \overline{B}) = A \cap (B \cup \overline{B})$$

$$= A$$

$$\therefore 1 - X_A(x) = \begin{cases} 1 & if \quad x \in \overline{A} \\ 0 & if \quad x \in A \end{cases}$$

$$= \begin{cases} 1 & if \quad x \in \overline{A} \\ 0 & if \quad x \notin \overline{A} \end{cases}$$

$$= X_{\overline{A}}(x)$$

P1.2 **Solution.**

$$X = [1,3], \ Y = [3,5]$$

$$\begin{aligned} &X + Y \\ &= [1+3, 3+5] \\ &= [4,8] \end{aligned}$$

$$\begin{aligned} &X - Y \\ &= [1-5, 3-3] \\ &= [-4, 0] \end{aligned}$$

$$\begin{aligned} &X \times Y \\ &= [\min\{1\times3, 1\times5, 3\times3, 3\times5\}, \max\{1\times3, 1\times5, 3\times3, 3\times5\}] \\ &= [3,15] \end{aligned}$$

$$X / Y$$

$$= X \cdot Y^{-1}$$

$$= [1,3] \cdot \left[\frac{1}{5}, \frac{1}{3}\right]$$

$$= \left[\min\left\{1\times\frac{1}{5}, 1\times\frac{1}{3}, 3\times\frac{1}{5}, 3\times\frac{1}{3}\right\}, \max\left\{1\times\frac{1}{5}, 1\times\frac{1}{3}, 3\times\frac{1}{5}, 3\times\frac{1}{3}\right\}\right]$$

$$= \left[\frac{1}{5}, 1\right]$$

$$\begin{aligned} &\max\{X, Y\} \\ &= [\max\{1,3\}, \max\{3,5\}] \\ &= [3,5] \end{aligned}$$

$$\begin{aligned} &\min\{X, Y\} \\ &= [\min\{1,3\}, \min\{3,5\}] \\ &= [1,3] \end{aligned}$$

P1.3 **Solution.**

$$\begin{aligned} &\det[R_I] \\ &= \begin{vmatrix} R_1 + R_2 & R_2 \\ R_2 & R_2 + R_3 \end{vmatrix} \\ &= (R_1 + R_2) \cdot (R_2 + R_3) - R_2 \cdot R_2 \end{aligned}$$

$$R_1 + R_2 = R_2 + R_3 = [0.8,1.2] + [0.8,1.2] = [1.6,2.4]$$
$$R_2 \times R_2 = [0.8,1.2] \times [0.8,1.2] = [0.64,1.44]$$
$$(R_1 + R_2) \times (R_2 + R_3) = [1.6,2.4] \times [1.6,2.4] = [2.56,5.76]$$
$$\det[R_I] = (R_1 + R_2) \times (R_2 + R_3) - R_2 \times R_2 = [1.12,5.12]$$

Notice that this routine calculation procedure needs 5 interval operations to get $\det[R_I]$, but

$$\det[R_I] = (R_1 + R_2) \times (R_2 + R_3) - R_2 \times R_2 = R_1(R_2 + R_3) + R_2 \times R_3$$

needs only 4 interval operations, since it cancels the term $R_2 \times R_2$ without computing it. This yields a better result: [1.92,4.32]. Using this approach, we have

$$\frac{adj[R_I]}{\det[R_I]} = \frac{\begin{bmatrix} R_2 + R_3 & -R_2 \\ -R_2 & R_1 + R_2 \end{bmatrix}}{[1.92,4.32]}$$

$$= \frac{\begin{bmatrix} [1.6,2.4] & [-1.2,-0.8] \\ [-1.2,-0.8] & [1.6,2.4] \end{bmatrix}}{[1.92,4.32]}$$

Therefore,

$$\begin{bmatrix} I_1 \\ I_2 \end{bmatrix} = \frac{adj[R_I]}{\det[R_I]} \begin{bmatrix} V_1 \\ V_2 \end{bmatrix}$$

$$= \begin{bmatrix} [1.6,2.4] & [-1.2,-0.8] \\ [-1.2,-0.8] & [1.6,2.4] \end{bmatrix} \begin{bmatrix} 1 \\ 0 \end{bmatrix} \times [1.92,4.32]^{-1}$$

$$= \frac{\begin{bmatrix} \frac{1.6}{4.32}, \frac{2.4}{1.92} \end{bmatrix}}{-\begin{bmatrix} \frac{0.8}{4.23}, \frac{1.2}{1.92} \end{bmatrix}}$$

$$= \begin{bmatrix} [0.37,1.25] \\ [-0.63,-0.19] \end{bmatrix}$$

P1.4   Solution.

(a) $Y = 1 + X + X^2 + X^3 + X^4 + X^5$
$$= 1 + X(1 + X(1 + X(1 + X(1 + X))))$$

$$= [1,1] + [2,3] ([1,1] + [2,3] ([1,1] + [2,3] ([1,1]$$
$$+ [2,3] ([1,1] + [2,3]))))$$
$$= [63,364]$$

In comparison, by a direct computation one obtains the same solution:

$$Y = 1 + X + X^2 + X^3 + X^4 + X^5$$
$$= 1 + [2,3] + [4,9] + [8,27] + [16,81] + [32,243]$$
$$= [63,364]$$

(b) $Y = \dfrac{X^3 - 1}{1 - X} = \dfrac{(X-1)(X^2 + X + 1)}{-(X-1)}$

$$= -1 - X - X^2$$
$$= [-6, -2] - [1,25]$$
$$= [-31, -3]$$

In comparison, by a direct computation one cannot complete the task:

$$Y = \frac{X^3 - 1}{1 - X} = \frac{[1,1] - [-1,1]^3}{[-1,1] - [1,1]} = \frac{[0,2]}{[-2,0]} \quad \text{(undefined)}$$

P1.5 **Solution.**

$X = [2,5]$ and $Y = [1,6]$

(a) For $z = x + y$

$$Z = [2,5] + [1,6] = [3,11]$$

$$X_{\bar{\alpha}} = [\alpha + 2, -2\alpha + 5]$$
$$Y_{\bar{\alpha}} = [3\alpha + 1, -2\alpha + 6]$$
$$\Rightarrow Z_{\bar{\alpha}} = [4\alpha + 3, -5\alpha + 11]$$

$$\therefore \mu_Z(z) = \begin{cases} \dfrac{z-3}{4} & 3 \le z \le 7 \\ \dfrac{11-z}{4} & 7 \le z \le 11 \end{cases}$$

(b) For $z = x - y$

$$Z = [2, 5] - [1, 6] = [-4, 4]$$

$$X_{\bar{\alpha}} = \left[\alpha + 2, \ -2\alpha + 5\right]$$
$$Y_{\bar{\alpha}} = \left[3\alpha + 1, \ -2\alpha + 6\right]$$
$$\Rightarrow Z_{\bar{\alpha}} = \left[3\alpha - 4, \ -5\alpha + 4\right]$$

$$\therefore \mu_Z(z) = \begin{cases} \dfrac{z+4}{3} & -4 \le z \le -1 \\[2ex] \dfrac{4-z}{5} & -1 \le z \le 4 \end{cases}$$

(c) For $z = x \ \vee \ y$

$$Z = \max\{x, y\} = [2,6]$$

$$X_{\bar{\alpha}} = \left[\alpha + 2, \ -2\alpha + 5\right]$$
$$Y_{\bar{\alpha}} = \left[3\alpha + 1, \ -2\alpha + 6\right]$$
$$\Rightarrow Z_{\bar{\alpha}} = \left[(\alpha + 2) \vee (3\alpha + 1), (-2\alpha + 5) \vee (-2\alpha + 6)\right]$$

Intersection: $\alpha + 2 = 3\alpha + 1 \Rightarrow \alpha = \dfrac{1}{2}$ but $-2\alpha + 5 = -2\alpha + 6$ has no intersection; therefore,

$$[z_1, z_2] = \begin{cases} [\alpha + 2, -2\alpha + 6] & 0 \le \alpha \le \dfrac{1}{2} \\[2ex] [3\alpha + 1, -2\alpha + 6] & \dfrac{1}{2} \le \alpha \le 1 \end{cases}$$

Finally,

$$\mu_Z(z) = \begin{cases} z - 2 & 2 \le z \le \dfrac{5}{2} \\[2ex] \dfrac{z-1}{3} & \dfrac{5}{2} \le z \le 4 \\[2ex] \dfrac{-z+6}{2} & 4 \le z \le 6 \end{cases}$$

(d) For $z = x \ \wedge \ y$

$$Z = \min\{x, y\} = [1,5]$$

$$X_{\bar{\alpha}} = \left[\alpha + 2, \ -2\alpha + 5\right]$$
$$Y_{\bar{\alpha}} = \left[3\alpha + 1, \ -2\alpha + 6\right]$$
$$\Rightarrow Z_{\bar{\alpha}} = \left[(\alpha + 2) \wedge (3\alpha + 1), (-2\alpha + 5) \wedge (-2\alpha + 6)\right]$$

Intersection: $\alpha + 2 = 3\alpha + 1 \Rightarrow \alpha = \dfrac{1}{2}$ but $-2\alpha + 5 = -2\alpha + 6$ has

no intersection; therefore,

$$[z_1, z_2] = \begin{cases} [3\alpha + 1, -2\alpha + 5] & 0 \le \alpha \le \dfrac{1}{2} \\[2ex] [\alpha + 2, -2\alpha + 5] & \dfrac{1}{2} \le \alpha \le 1 \end{cases}$$

Finally,

$$\mu_Z(z) = \begin{cases} \dfrac{z-1}{3} & 1 \le z \le \dfrac{5}{2} \\[2ex] z - 2 & \dfrac{5}{2} \le z \le 3 \\[2ex] \dfrac{-z+5}{2} & 3 \le z \le 5 \end{cases}$$

CHAPTER 2

P2.1   Solution.

For instance in the modus ponens, the inference rule is

$$(a \wedge (a \Rightarrow b)) = b$$

In terms of membership values, this is equivalent to

IF $\mu(a) = 0.9$

$$\begin{aligned} \text{AND } \mu(a \Rightarrow b) &= \min\{\, 1, 1 + \mu(b) - \mu(a) \,\} \\ &= \min\{\, 1, 1 + \mu(b) - 0.9 \,\} \\ &= \min\{\, 1, 0.1 + \mu(b) \,\} \\ &= 0.3 \end{aligned}$$

THEN $\mu(b) = 0.3 - 0.1 = 0.2$

P2.2   Solution.

From Example 2.4, the membership function for the fuzzy description "small" of A = {1, 2, 3, 4} is as follows:

$$\mu(a) = \begin{cases} 1.0 & a = 1, \\ 0.7 & a = 2, \\ 0.3 & a = 3, \\ 0.0 & a = 4, \end{cases}$$

and the fuzzy relation R between two members in A, meaning "approximately equal," is defined by the following table:

	1	2	3	4
1	1.0	0.5	0.0	0.0
2	0.5	1.0	0.5	0.0
3	0.0	0.5	1.0	0.5
4	0.0	0.0	0.5	1.0

R :

$$\mu_B(b) = \max_{a \in A} \{\min\{\mu_A(a), \mu_{\tilde{R}}(a,b)\}\}, \quad b \in B = A.$$

For $b = 1$,

$$\begin{aligned}
\mu_B(1) &= \max_{a \in A} \{\min\{\mu_A(a), \mu_{\tilde{R}}(a,1)\}\} \\
&= \max\{ \min\{1.0,1.0), \min\{0.7,0.5\}, \min\{0.3,0.0\}, \\
&\qquad\qquad \min\{0.0,0.0\} \} \\
&= \max\{1.0, 0.5, 0.0, 0.0\} \\
&= 1.0
\end{aligned}$$

For $b = 3$,

$$\begin{aligned}
\mu_B(3) &= \max_{a \in A} \{\min\{\mu_A(a), \mu_{\tilde{R}}(a,3)\}\} \\
&= \max\{ \min\{1.0,0.0\}, \min\{0.7,0.5\}, \min\{0.3,1.0\}, \\
&\qquad\qquad \min\{0.0,0.5\} \} \\
&= \max\{ 0.0, 0.5, 0.3, 0.0 \} \\
&= 0.5
\end{aligned}$$

For $b = 4$,

$$\begin{aligned}
\mu_B(4) &= \max_{a \in A} \{\min\{\mu_A(a), \mu_{\tilde{R}}(a,3)\}\} \\
&= \max\{ \min\{1.0,0.0\}, \min\{0.7,0.0\}, \min\{0.3,0.5\}, \\
&\qquad\qquad \min\{0.0,1.0\} \} \\
&= \max\{ 0.0, 0.0, 0.3, 0.0 \} \\
&= 0.3
\end{aligned}$$

Hence, the results are verified.

P2.3 **Solution.**

$$\mu_B(5) = \max_{a \in A} \left\{ \min \left\{ \mu_A(a), \mu_{\tilde{R}}(a,5) \right\} \right\}$$
$$= \max\{ \min\{1.0,0.5\}, \min\{0.5,1.0\}, \min\{0.1,0.5\}, \min\{0.0,0.1\} \}$$
$$= \max\{ 0.5, 0.5, 0.1, 0.1 \}$$
$$= 0.5$$

P2.4 **Solution.**

Premise	The car is not affordable
Implication	If the price is cheap then the car is affordable
Conclusion	The price is not cheap

P2.5 **Solution.**

IF $\mu(\bar{a}) = 0.7$

$$\text{AND } \mu(\bar{a} \Rightarrow b) = \min\{ 1, 1 + \mu(b) - \mu(\bar{a}) \}$$
$$= \min\{ 1, \mu(b) + 0.3 \}$$
$$= 0.8$$

THEN $\mu(b) = 0.8 - 0.3 = 0.5$

P2.6 **Solution.**

(a) $\mu(a \Rightarrow b) = \min\{ 1, 1 + \mu(b) - \mu(a) \}$
$$= \min\{ 1, 0.2 \}$$
$$= 0.2$$
<u>Yes</u>, this result is correct.

(b) $\mu(a \Rightarrow b) = \min\{ 1, 1 + \mu(b) - \mu(a) \}$
$$= \min\{ 1, 1 + 0.9 - 0.2 \}$$
$$= \min\{ 1, 1.7 \}$$
$$= 1.0$$

No, this result is not meaningful. The main reason is that the inference statement is an "overconfident" claim based on little information with very low confidence.

P2.7   **Solution.**

IF b is $\overline{B}$ AND c is $\overline{C}$ THEN a is $\overline{A}$

CHAPTER 3

P3.1   **Solution.**

This is a take-home project assignment.

P.3.2   **Solution.**

This is a take-home project assignment.

CHAPTER 4

P4.1   **Solution.**

A rule base can be written based on the following table:

IF	x1		x2	THEN	y
	NL	AND	PS		PM
	PS	AND	NL		PM
	NM				PM
			NM		PM
	PM				PM
			PM		PM
Otherwise					PL

P4.2   **Solution.**

R^1:   IF $a > b$ THEN $y = ab$
R^2:   IF $a < b$ THEN $y = a + b$
R^3:   IF $a \in R$ AND $b \in R$ AND $a = b$ THEN $y = 0$

Otherwise, y = ∞

(If a or b is not real, then they cannot compare each other by < or >, so this case belongs to "Otherwise.")

P4.3 Solution.

(a) $\mu(a \Rightarrow b)$ = min{ 1, 1 + $\mu(b)$ − $\mu(a)$ }

$\Rightarrow$ 0.2 = min{ 1, 1 + $\mu(b)$ − 0.9 }

$\Rightarrow$ 0.2 = min{ 1, 0.1 + $\mu(b)$ }

$\Rightarrow$ 0.2 = 0.1 + $\mu(b)$

$\Rightarrow$ $\mu(b)$ = 0.1

(b) $\mu(a \Rightarrow b)$ = min{ 1, 1 + $\mu(b)$ − $\mu(a)$ }

= min{ 1, 1 + $\mu(b)$ − (1 − $\mu(\bar{a})$) }

$\Rightarrow$ 0.8 = min{ $\mu(b)$ + 0.8 }

$\Rightarrow$ $\mu(b)$ = 0

P4.4 Solution.

Follow Example 1.11 and Example 1.12 in Chapter 1.

P4.5 Solution.

(a) There should not be AND in the conclusion part.

"IF a is A THEN b is $\bar{B}$ "

"IF a is A THEN $\bar{c}$ is C "

(b) The rule base has two rules:

"IF a is A THEN c is $\bar{C}$ "

"IF b is B THEN c is $\bar{C}$ "

(c) "If b is B AND c is $\bar{C}$ THEN a is A"

CHAPTER 5

P5.2 Solution.

Here, clockwise rotation is considered to be positive and anti-clockwise is negative.

Angle θ	*x*-Position	Steering angle *u*
AB: Right	AM: Right much	NB: Negative big
AC: Right near center	AR: Right	NM: Negative medium
CE: Center	AV: Right near vertical axis	NS: Negative small
BC: Left near center	ZO: Zero	ZE: Zero
EE: Left	BV: Left near vertical axis	PS: Positive small
	BR: Left	PM: Positive medium
	BM: Left much	PB: Positive big

The membership function of angle θ is as shown below:

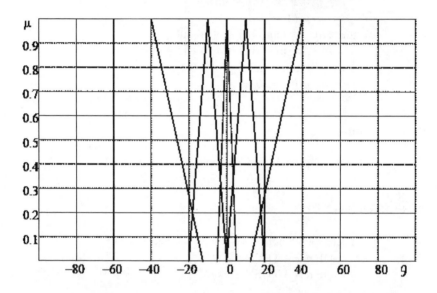

The input variables are the truck angle θ and the horizontal position coordinate x, while the output variable is the steering angle (signal) u. The variable are pre-assigned as

$$-100 \leq x \leq 100$$
$$-180° \leq \theta \leq 180°$$
$$-30° \leq u \leq 30°$$

Here, clockwise rotation is considered to be positive and anti-clockwise is negative.

Angle θ	*x*-Position	Steering angle *u*
AB: Right	AM: Right much	NB: Negative big
AC: Right near center	AR: Right	NM: Negative medium
CE: Center	AV: Right near vertical axis	NS: Negative small
BC: Left near center	ZO: Zero	ZE: Zero
EE: Left	BV: Left near vertical axis	PS: Positive small
	BR: Left	PM: Positive medium
	BM: Left much	PB: Positive big

The membership function of angle θ is as shown below:

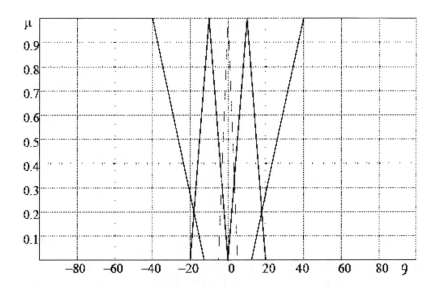

The membership function of the *x* position is as shown below:

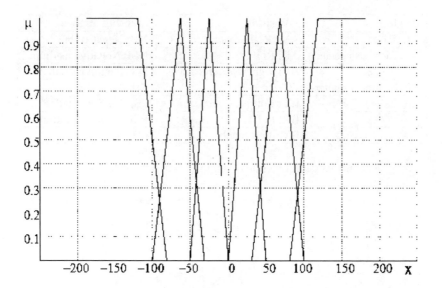

The membership function for the control signal u is as shown below:

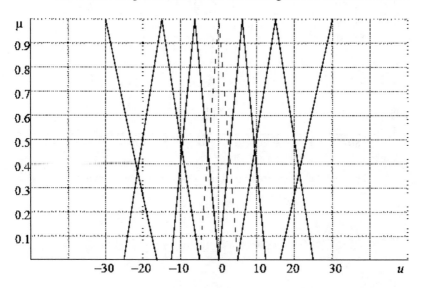

The rule base is summarized as follows:

R: IF x is X AND θ is Θ THEN u is U

	BE	BC	CE	AC	AB
BM	PB	PB	PM	PM	PS
BR	PB	PB	PM	PS	NS
BV	PB	PM	PS	NS	NM
ZO	PM	PM	ZE	NM	NM
AV	PM	PS	NS	NM	NB
AR	PS	NS	NM	NB	NB
AM	NS	NM	NM	NB	NB

CHAPTER 6

P6.1 **Solution.**

It can be seen from the figure that

$$U(s) = K_P E(s) + K_D s E(s)$$
$$E(s) = R(s) - Y(s)$$
$$Y(s) = \frac{1}{as+b} U(s)$$

so that

$$Y(s) = H(s)R(s) = \frac{K_P + K_D s}{(a+K_D)s + (b+K_P)} R(s)$$

The transfer function $H(s)$ of the overall feedback control system has one pole:

$$s_1 = \frac{-(b+K_P)}{a+K_D}$$

This pole is real and is required to be negative for stability. Because $a < 0$, one has

$$K_P > -b \quad \text{and} \quad K_D > -a$$

Next, using the Terminal Value Theorem of Laplace Transforms, one has

$$\lim_{t \to \infty} e(t) = \lim_{|s| \to 0} sE(s)$$

$$= \lim_{|s| \to 0} s[1 - H(s)]R(s)$$

$$= \lim_{|s| \to 0} s \cdot \frac{as + b}{(a + K_D)s + (b + K_P)} \cdot \frac{r}{s}$$

$$= \frac{br}{b + K_P}$$

which is not zero unless $b = 0$ or $r = 0$. This means that the PD controller cannot eliminate the steady-state tracking error, even for this first-order linear plant.

P6.2   **Solution.**

It can be seen from the figure that

$$U(s) = K_P E(s) + \frac{1}{s} K_I E(s)$$

$$E(s) = R(s) - Y(s)$$

$$Y(s) = \frac{1}{as^2 + b} U(s)$$

so that

$$Y(s) = H(s)R(s) = \frac{sK_P + K_I}{as^3 + (b + K_P)s + K_I} R(s)$$

Suppose that the three poles of the transfer function $H(s)$ are s_1, s_2 and s_3. Then

$$as^3 + (b + K_P)s + K_I = a(s - s_1)(s - s_2)(s - s_3)$$

$$= a[s^3 - (s_1 + s_2 + s_3)s^2 + (s_1 s_2 + s_2 s_3 + s_3 s_1)s - s_1 s_2 s_3]$$

A comparison of their coefficients shows that $s_1 + s_2 + s_3 \equiv 0$, which implies that the three poles cannot all have negative real parts. Thus, the system stability cannot be guaranteed by any choice of the control gains.

Next, using the Terminal Value Theorem of Laplace Transforms, one has

$$
\begin{aligned}
\lim_{t \to \infty} e(t) &= \lim_{|s| \to 0} sE(s) \\
&= \lim_{|s| \to 0} s[1 - H(s)]R(s) \\
&= \lim_{|s| \to 0} s \cdot \frac{as^3 + bs}{as^3 + (b + K_P)s + K_I)} \cdot \frac{r}{s} \\
&= 0
\end{aligned}
$$

which implies that the set-point tracking can always be achieved no matter what control gains are used.

This particular example shows that, sometimes, the control gains of a PID controller may not provide much freedom for tracking control and stability warranty purposes.

P6.3   Solution.

Using the bilinear transform in the continuous-time D controller, one obtains

$$
s K_D \to \frac{2}{T} \frac{z-1}{z+1} K_D = \frac{2}{T} \frac{1-z^{-1}}{1+z^{-1}} K_D,
$$

$$
U(s) = s K_D E(s) \to \tilde{U}(z) = \frac{2}{T} \frac{1-z^{-1}}{1+z^{-1}} K_D \tilde{E}(z).
$$

Let $\tilde{K}_D = \dfrac{2}{T} K_D.$ Then,

$$
(1 + z^{-1}) \tilde{U}(z) = \tilde{K}_D (1 - z^{-1}) \tilde{E}(z),
$$

so that the inverse z-transform gives

$$
u(nT) = - u(nT{-}T) + \tilde{K}_D [e(nT) - e(nT{-}T)].
$$

This digital D controller can be implemented as shown in the following figure:

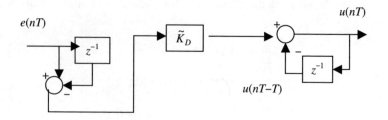

The digital D controller

CHAPTER 7

P7.1   Solution.

Assume the apple initially has a height of h_0. Then it follows from Newton's law that

$$\varepsilon_{\text{fall}}(t) = h_0 - \frac{1}{2} g t^2.$$

P7.2   Solution.

(i) The temperature *decreases* from 10 to 0. Boundary conditions are given by

$$\varepsilon_{\text{decrease}}(0) = 10, \qquad \varepsilon_{\text{decrease}}(T_w) = 0$$

$$\Rightarrow \quad h_0 + ke^0 = 10, \quad h_0 + ke^{T_w} = 0$$

$$\Rightarrow \quad h_0 + k = 10, \qquad h_0 + ke^{T_w} = 0$$

$$\Rightarrow \quad k = \frac{10}{1 - e^{T_w}}, \quad h_0 = \frac{10e^{T_w}}{1 - e^{T_w}}$$

Therefore,

$$\varepsilon_{\text{become}}(t) = \frac{10e^{T_w}}{1-e^{T_w}} + \frac{10}{1-e^{T_w}}e^{-t}$$

(ii) The temperature *decreases* from *around ten* to *around zero.* Denote the membership function of a triangular fuzzy number *close to a* with support 2 as $[a-1, a, a+1]$. Then $\varepsilon_{\text{decrease}}$ can be determined by three functions in the form of $\varepsilon_{\text{decrease}}(t) - 1$, $\varepsilon_{\text{decrease}}(t)$, and $\varepsilon_{\text{decrease}}(t) + 1$, where $\varepsilon_{\text{decrease}}(t)$ is the same as above.

P7.3　**Solution.**

    (i) *increase fast:* $\varepsilon_{\text{increasefast}}(t) = fast \circ e^t = e^{at}, \text{a} > 1.$

    (ii) *increase slowly:* $\varepsilon_{\text{increaseslowly}}(t) = slowly \circ e^t = e^{at}, \text{a} < 1.$

    (iii) *decrease:* $\varepsilon_{\text{decrease}}(t) = reversely \circ e^t = e^{-t}.$

P7.4　**Solution.**

    (i) $S = \dfrac{\displaystyle\int_0^2 e^{-2t}\,dt}{\displaystyle\int_0^2 e^{-t}\,dt} = \dfrac{\dfrac{1}{2} - \dfrac{e^{-4}}{2}}{1-e^{-2}} = \dfrac{1}{2}\dfrac{1-e^{-4}}{1-e^{-2}}$

    (ii) $S = \dfrac{\displaystyle\int_0^2 e^{-t}\,dt}{\displaystyle\int_0^2 e^t\,dt} = \dfrac{1-e^{-2}}{e^2-1}$

P7.5　**Solution.**

    (i) The canonical computational verbs in *become* are *become* (*here*, museum) or *become* (*not near* museum, museum)

(ii) The canonical computational verbs in *become* are *become* (current speed, bigger than current speed) or *become* (slow, fast)

P7.6   Solution.

(i) The linear interpretation of V_{e2}. Note that the corresponding waveform of $|e(t)|$, $\varepsilon_2(t)$, passes points $(0, 0.5)$, $(T_w/2, 0.375)$ and $(T_w, 0.25)$. Therefore, the linear interpretation of Ve2 is

$$e_2(t) = 0.5 - \frac{t}{4T_w}, \quad t \in [0, T_w]$$

(ii) The linear interpretation of V_{e3}. Note that the corresponding waveform of $|e(t)|$, $\varepsilon_3(t)$, passes points $(0, 0.25)$, $(T_w/2, 0.125)$ and $(T_w, 0)$. Therefore, the linear interpretation of V_{e3} is

$$e_3(t) = 0.25 - \frac{t}{4T_w}, \quad t \in [0, T_w]$$

(iii) The linear interpretation of V_{e4}. Note that the corresponding waveform of $|e(t)|$, $\varepsilon_4(t)$, passes points $(0, 0)$, $(T_w/2, 0.125)$ and $(T_w, 0.25)$. Therefore, the linear interpretation of V_{e4} is

$$e_4(t) = \frac{t}{4T_w}, \quad t \in [0, T_w]$$

(iv) The linear interpretation of V_{e5}. Note that the corresponding waveform of $|e(t)|$, $\varepsilon_5(t)$, passes points $(0, 0.25)$, $(T_w/2, 0.375)$ and $(T_w, 0.5)$. Therefore, the linear interpretation of V_{e5} is

$$e_5(t) = 0.25 + \frac{t}{4T_w}, \quad t \in [0, T_w]$$

(v) The linear interpretation of V_{e6}. Note that the corresponding waveform of $|e(t)|$, $\varepsilon_6(t)$, passes points $(0, 0.5)$, $(T_w/2, 0.75)$ and $(T_w, 1)$. Therefore, the linear interpretation of V_{e6} is

$$e_6(t) = 0.5 + \frac{t}{2T_w}, \quad t \in [0, T_w]$$

(vi) The linear interpretation of V_{e7}. Note that the corresponding waveform of $|e(t)|$, $\varepsilon_7(t)$, passes points $(0, 0)$, $(T_w/2, 0)$ and $(T_w, 0)$. Therefore, the linear interpretation of V_{e7} is

$$e_6(t) = 0, \qquad t \in [0, T_w]$$

P7.7   Solution.

(i) $\xi_{V_{e2}}^{V_e} = \dfrac{0}{V_{e1}} + \dfrac{1}{V_{e2}} + \dfrac{0}{V_{e3}} + \dfrac{0}{V_{e4}} + \dfrac{1}{9V_{e5}} + \dfrac{0}{V_{e6}} + \dfrac{0}{V_{e7}}.$

(ii) $\xi_{V_{e3}}^{V_e} = \dfrac{0}{V_{e1}} + \dfrac{0}{V_{e2}} + \dfrac{1}{V_{e3}} + \dfrac{1}{9V_{e4}} + \dfrac{0}{V_{e5}} + \dfrac{0}{V_{e6}} + \dfrac{1}{9V_{e7}}.$

(iii) $\xi_{V_{e4}}^{V_e} = \dfrac{0}{V_{e1}} + \dfrac{0}{V_{e2}} + \dfrac{1}{9V_{e3}} + \dfrac{1}{V_{e4}} + \dfrac{0}{V_{e5}} + \dfrac{0}{V_{e6}} + \dfrac{1}{9V_{e7}}.$

(iv) $\xi_{V_{e5}}^{V_e} = \dfrac{0}{V_{e1}} + \dfrac{1}{9V_{e2}} + \dfrac{0}{V_{e3}} + \dfrac{0}{V_{e4}} + \dfrac{1}{V_{e5}} + \dfrac{0}{V_{e6}} + \dfrac{0}{V_{e7}}.$

(v) $\xi_{V_{e6}}^{V_e} = \dfrac{1}{9V_{e1}} + \dfrac{0}{V_{e2}} + \dfrac{0}{V_{e3}} + \dfrac{0}{V_{e4}} + \dfrac{0}{V_{e5}} + \dfrac{1}{V_{e6}} + \dfrac{0}{V_{e7}}.$

(vi) $\xi_{V_{e7}}^{V_e} = \dfrac{0}{V_{e1}} + \dfrac{0}{V_{e2}} + \dfrac{1}{9V_{e3}} + \dfrac{1}{9V_{e4}} + \dfrac{0}{V_{e5}} + \dfrac{0}{V_{e6}} + \dfrac{1}{V_{e7}}.$

Index